KV-080-891

THOMOND RESOURCE CENTRE

Plassey, Limerick

Telephone (061) 334488

Acc. No. 2627507

PLEASE RETURN BY LATEST DATE SHOWN

0 4 DEC 1991		
- 9 SEP 1992		
	0 2 NOV 1996	

2627507

ENJOYIN
ELECTRO

Owen Bis

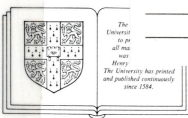

The
Universit
to pr
all ma
was
Henry
The University has printed
and published continuously
since 1584.

CAMBRIDGE UNIVERSITY PRESS

Cambridge
New York   Port Chester
Melbourne   Sydney

Contents

Published by the Press Syndicate of the University of Cambridge
The Pitt Building, Trumpington Street, Cambridge CB2 1RP
32 East 57th Street, New York, NY 10022, USA
10 Stamford Road, Oakleigh, Melbourne 3166, Australia

© Cambridge University Press 1983

First published 1983
Third printing 1989

Printed in Great Britain at the University Press, Cambridge

British Library cataloguing in publication data

Bishop, Owen
  Enjoying electronics.
  1. Electronics
  I. Title
  537.5 TK7816

ISBN 0 521 28773 1

PLASSEY CAMPUS LIBRARY	
TCE FUND	
BRN	81771-1
LOC	TRC
CLASS	537.5

Illustrations by Colin King and
Jeff Edwards, Marlborough Design

What is electronics?

All substances are made of atoms. Atoms are too small to be seen.

Atoms are so small that an apple is made from about 10 000 000 000 000 000 000 000 atoms.

There are lots of different kinds of atoms, but they are all built in the same way. They all have a nucleus. There is a cloud of electrons around the nucleus.

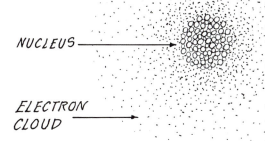

NUCLEUS ⟶

ELECTRON CLOUD ⟶

1 000 000 000 000 000 000 000 000 000 electrons weigh the same as a 1 kilogram bag of sugar

Different kinds of atom have different sizes of nucleus and different numbers of electrons.

The nucleus of an atom carries positive electric charges. The electrons each carry a negative electric charge. Electrons are very small and very light in weight, but there are lots of them in any small piece of substance.

Electronics uses electrons to do useful jobs. This book will help you to understand how we use electrons. It will show you how to control electrons and make them do what you want.

1 Making a circuit

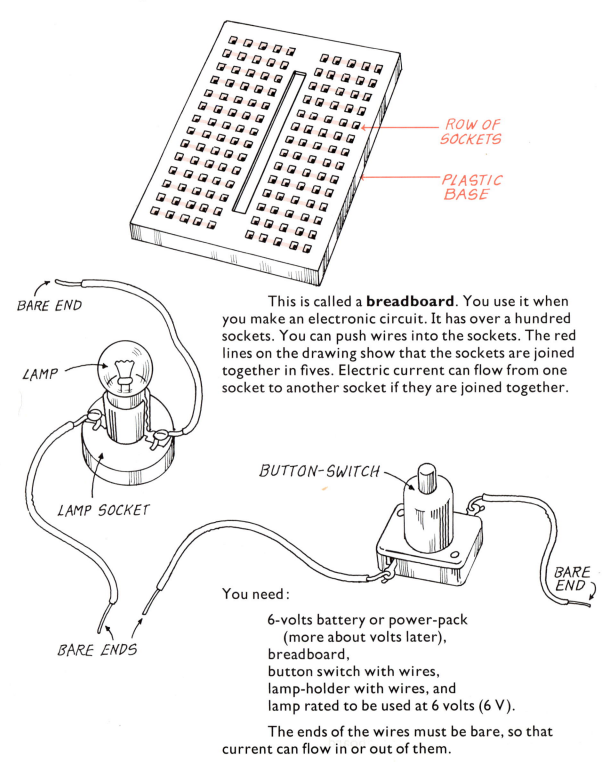

ROW OF SOCKETS

PLASTIC BASE

BARE END

LAMP

LAMP SOCKET

BARE ENDS

BUTTON-SWITCH

BARE END

This is called a **breadboard**. You use it when you make an electronic circuit. It has over a hundred sockets. You can push wires into the sockets. The red lines on the drawing show that the sockets are joined together in fives. Electric current can flow from one socket to another socket if they are joined together.

You need:

    6-volts battery or power-pack
      (more about volts later),
    breadboard,
    button switch with wires,
    lamp-holder with wires, and
    lamp rated to be used at 6 volts (6 V).

The ends of the wires must be bare, so that current can flow in or out of them.

TRY THIS CIRCUIT

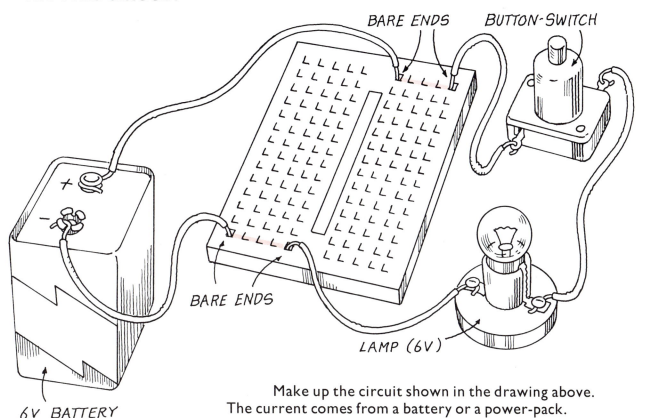

BARE ENDS BUTTON-SWITCH

BARE ENDS

LAMP (6V)

6V BATTERY
OR POWER PACK

Make up the circuit shown in the drawing above. The current comes from a battery or a power-pack.

When you have plugged all the wires into the breadboard, press the button-switch. The lamp lights. An electric current is flowing from the battery, along the wire to the breadboard, on through the switch, through the lamp, to the breadboard again, and finally back to the battery. It goes all the way round and back again – a complete **circuit**.

Pressing the button-switch **closes** the circuit. Then current can flow. When the button-switch is not pressed, the circuit is **open**. Then current cannot flow.

Here is an easier way to draw the circuit. On the left is a drawing of the circuit you have just built on the breadboard.

Can you see that it is a closed circuit when the button is pressed? Start at the positive (+) side of the battery and follow round the circuit until you come to the battery again. Check that this drawing shows the parts in the same order as the circuit on the breadboard.

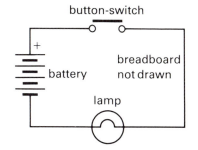

A 6-volts battery is usually made from four cells joined together. The sign for the battery shows these four cells.

3

2 Two lamps together

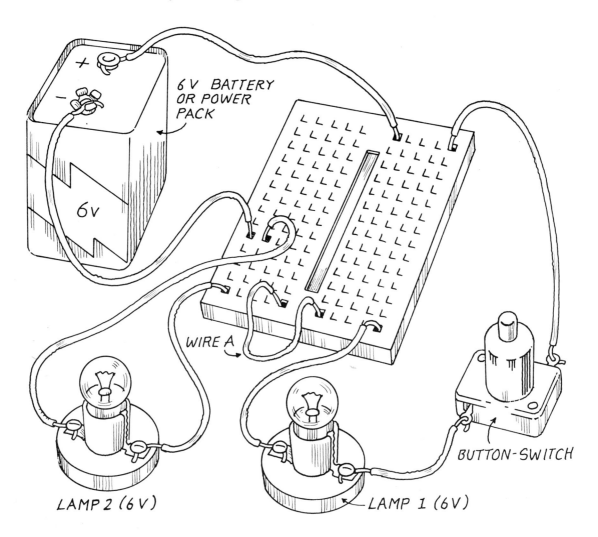

6 V BATTERY OR POWER PACK

6v

WIRE A

LAMP 2 (6V)

LAMP 1 (6V)

BUTTON-SWITCH

If you have forgotten how bright the one lamp was, wire up the circuit on page 3 again and press the button.

Make this circuit. Press the button-switch. Do the lamps shine as brightly as the lamp in the circuit on page 3?

Take away wire A and press the button again. What happens? Why? Is the circuit open or closed when you take away wire A?

Put wire A back in its sockets again. Unscrew one of the lamps from its holder. Press the button. What happens now? Why?

Replace the lamp in its holder. Press the button again. The lamps should both light now.

The current flows through lamp 1. Then it flows through lamp 2. We say that the lamps are in series. 'Series' means 'one after the other', like a series of television programmes.

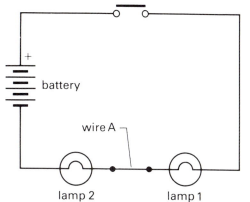

Now put in an extra wire, wire B. The drawing below shows you where to put it. Press the button. What happens now?

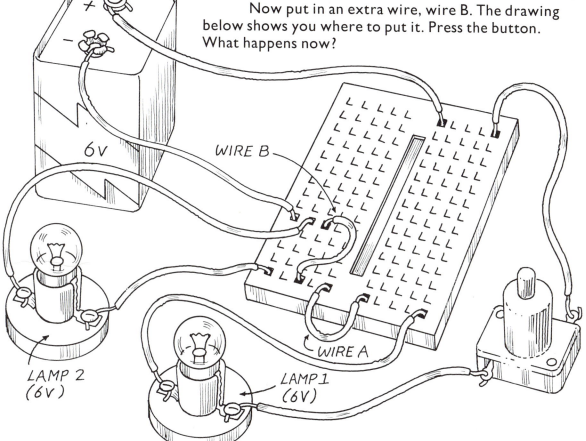

Can you see how wire B by-passes lamp 2?

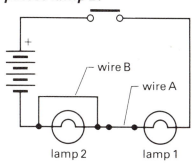

It is easy for the current to flow along ordinary wires. It is very much harder for it to flow along the very thin wire inside the lamps. So most of the current takes the easy way. Most of it goes through wire B and not through lamp 2. Wire B has made a **short-circuit** across lamp 2. A short-circuit is a kind of short cut. A short-circuit is often called a 'short'.

5

3 Lamps in parallel

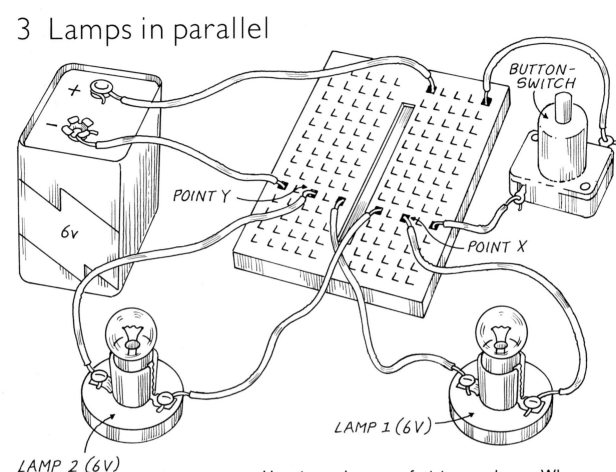

POINT Y

BUTTON-SWITCH

6v

POINT X

LAMP 1 (6V)

LAMP 2 (6V)

Here is another way of wiring two lamps. When the current reaches point X, which way can it go next? Press the button. Does the current go only one way or both ways? Do the lamps shine as brightly as the lamp in the circuit on page 3?

At point X the current splits into two. If the lamps are alike, equal currents flow to each lamp. These currents join again at point Y.

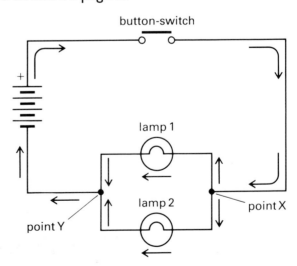

button-switch

+

lamp 1

point Y

lamp 2

point X

The lamps share the current as it passes around the circuit. We say that the lamps are wired **in parallel**.

If lamps are wired in parallel, we can put **two** switches into the circuit. There is one switch in each parallel section of circuit. One switch controls one lamp, the other switch controls the other lamp. See if you can wire up a circuit like this one on your breadboard.

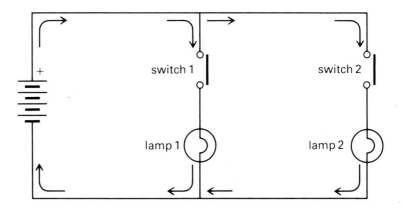

If you were an electrician wiring up lamps for lighting a house, would you connect them:

in series – as on page 4?

or in parallel – as on page 6?

(a)   (b)

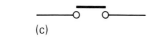

(c)

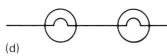

(d)

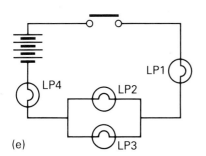

(e)

REVISION QUESTIONS

1 Which are the two main parts of an atom?

2 Which kind of electric charge is carried by an electron?

3 Look at the symbols (a), (b) and (c), drawn on the left. Say what each of them means.

4 What word is used to describe two lamps wired as in drawing (d) on the left?

5 When two lamps are wired in parallel, are they brighter or less bright than a single lamp, or is there no difference? (Assume that all lamps are of the same kind, powered by the same battery.)

6 Say what you think will happen in the circuit (e) on the left when the button-switch is pressed.

4 Electrons and currents

Electric currents flow through wires very easily. Wires are made of metal (usually copper). There are lots of electrons in the wire. The electrons come from the atoms of metal, but are able to move around in the spaces between the atoms. The electrons are **charge carriers**. When the electrons flow in the wire there is an electric current.

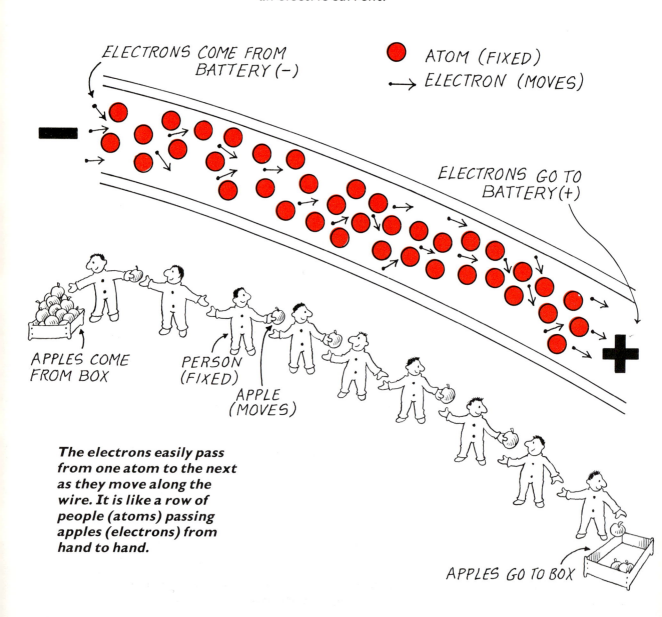

ELECTRONS COME FROM BATTERY (−)

● ATOM (FIXED)
⟶ ELECTRON (MOVES)

ELECTRONS GO TO BATTERY (+)

APPLES COME FROM BOX

PERSON (FIXED)

APPLE (MOVES)

The electrons easily pass from one atom to the next as they move along the wire. It is like a row of people (atoms) passing apples (electrons) from hand to hand.

APPLES GO TO BOX

CONDUCTORS AND NON-CONDUCTORS

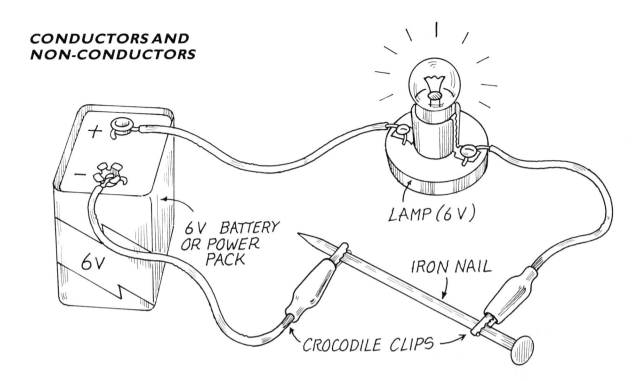

6 V BATTERY OR POWER PACK

6V

LAMP (6 V)

IRON NAIL

CROCODILE CLIPS

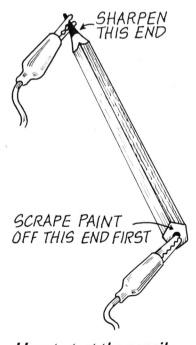

SHARPEN THIS END

SCRAPE PAINT OFF THIS END FIRST

How to test the pencil 'lead'. It is not really made of lead metal, but of carbon, which is not a metal.

Materials which carry current easily are called **conductors**. Those which do not carry current are called **non-conductors** or **insulators**. Test different materials to see if they are conductors or non-conductors.

First test iron. Fix the clips to the ends of an iron nail. Does the lamp shine or not? What does this tell you? Remove the clips from the nail as soon as the test is finished. If the lamp shines, it tells you that the circuit is closed. The current can flow through the iron nail. Iron is a conductor.

Now test these and any other materials you can think of:

strip of paper	piece of brick
'copper' coin	'silver' coin
plastic rod	piece of glass
strip of aluminium foil	the 'lead' of a pencil
piece of stone	piece of wood

Which of these are conductors? Make a list. Then write out a list of non-conductors.

5 Charge carriers

The atoms of conducting materials let their electrons go free very easily. These free electrons can then carry electric charge. Electrons carry negative charge, so we call them negative charge carriers.

The atoms of non-conducting materials do not let their electrons go free. There are no free electrons to carry charge. No current can flow.

There are no apple carriers here. The people are keeping their hands in their pockets!

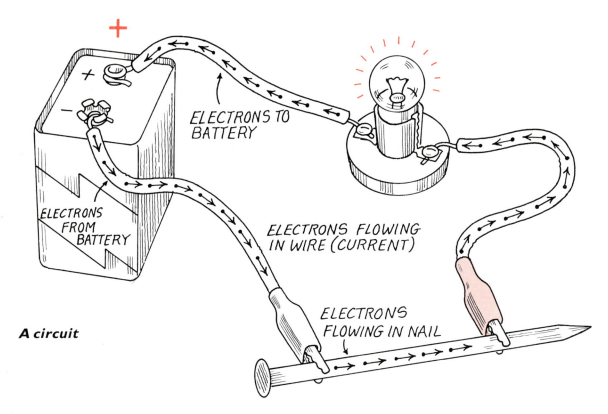

ELECTRONS TO BATTERY

ELECTRONS FROM BATTERY

ELECTRONS FLOWING IN WIRE (CURRENT)

ELECTRONS FLOWING IN NAIL

A circuit

This is how the current flows in a circuit. The charge carriers are electrons, carrying negative charge from the negative terminal of the battery to the positive terminal.

In some kinds of conducting material there may be other kinds of charge carrier, carrying positive charge. See what happens in the two kinds of material when a current flows.

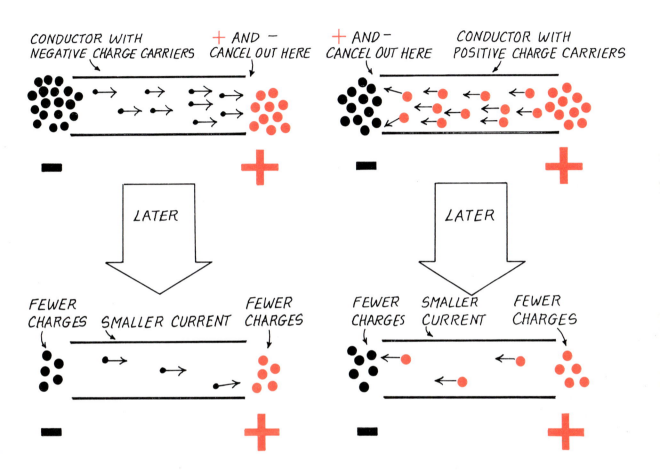

Current flows for as long as there is positive charge at one end of the material and negative charge at the other. One material has negative charge carriers (electrons) and the other has positive charge carriers, but the result is the same for both materials:

Positive gets less positive.
Negative gets less negative.

The result is a transfer of charge, from positive to negative.

The rate at which the charge is transferred is called the current.

6 Currents

A current of 1 amp is enough to make four 60-watt house lamps shine brightly.

The current going through the meter on the right is 64 milliamps.

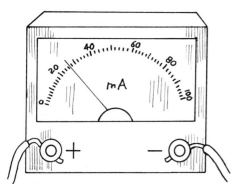

What is the current through this meter?

Current is measured in **amperes**. These are usually called **amps** for short. One amp (1A) is quite a large amount of current. The current in your circuits are usually much less than one amp. We often measure small currents in **milliamps**. A thousand milliamps (1000 mA) equals one amp.

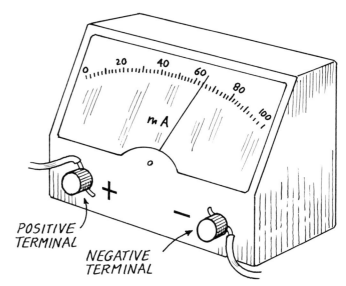

POSITIVE TERMINAL

NEGATIVE TERMINAL

This is an **ammeter**. It is used for measuring current. It has two terminals, positive and negative. It must be connected in a circuit the right way round. This ammeter measures currents up to 100 milliamps. It must not be used for larger currents.

This test is like the one on page 9. Test some of the materials you tested before. This time, the meter measures the current through the material.

Write your results in a table like this one:

MEASURING CURRENTS

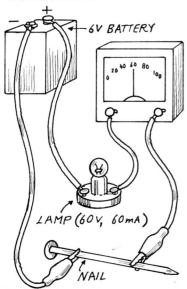

6V BATTERY

LAMP (6OV, 60mA)

NAIL

MATERIAL TESTED	CURRENT (MILLIAMPS)	BRIGHTNESS OF THE LAMP
IRON		
COPPER WIRE		
CARBON (PENCIL)		
PAPER		

Try the materials listed above, and a few more as well. Now try water from the tap. Is tap water a conductor?

This is how to measure currents through water.

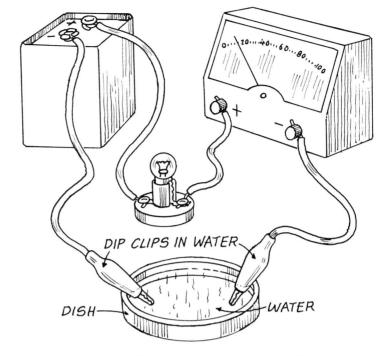

DIP CLIPS IN WATER

DISH — WATER

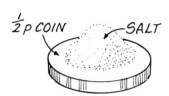

½ p COIN — SALT

Next add a little salt to the water. Watch the needle of the meter as the salt dissolves in the water. Which is the better conductor, tap water or salt water?

SUMMING UP

Your results can be explained like this:

1 Metals are good conductors, with lots of free electrons to carry charge. The lamp shines brightly.

2 Non-conductors have no charge carriers, so there is no current and the lamp does not shine.

3 Carbon does not have as many charge carriers as metals. The current is enough to light the lamp, but not brightly.

4 Water has very few charge carriers, giving a very small current which is not enough to light the lamp.

5 In salt water, the dissolved salt provides more charge carriers. A bigger current flows, but it is still not enough to light the lamp.

MORE CHARGE CARRIERS

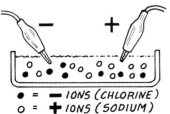

● = − IONS (CHLORINE)
○ = + IONS (SODIUM)

When salt (sodium chloride) dissolves in water, its molecules split into sodium atoms and chlorine atoms. The sodium atoms each give an electron to each of the chlorine atoms. The atoms became charged atoms, or **ions**. These ions act as charge carriers. They carry current through the salt solution. Which way(s) do the ions move?

13

7 Current and heat

This is a drawing of electrons flowing through a wire.

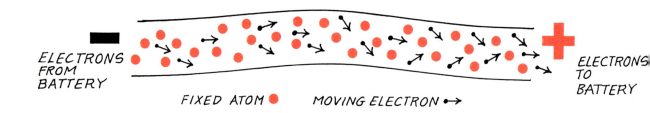

ELECTRONS FROM BATTERY

ELECTRONS TO BATTERY

FIXED ATOM ●   MOVING ELECTRON ➙

This is one way we use the heat made by an electric current.

The electrons are moving. They have energy, given to them by the battery. They move from atom to atom, in the space between the atoms. Some of the energy of the electrons is given to the atoms. As the atoms gain energy, the wire becomes warm.

If the current is large, the wire may get so warm that it glows red hot.

A large current through a thin wire makes the wire so hot that it becomes white hot and shines brightly. The wire in a lamp is made from a metal called **tungsten**. This can be made very hot without it melting. We put a glass bulb around the wire to stop the air from getting to it and burning it away.

LAMP

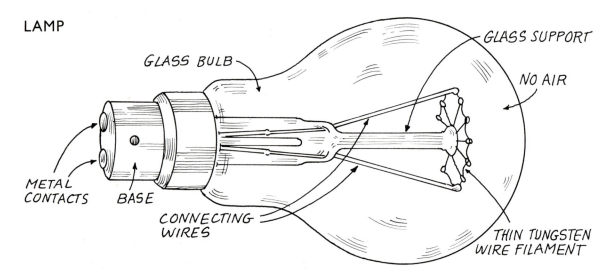

GLASS SUPPORT

GLASS BULB

NO AIR

METAL CONTACTS

BASE

CONNECTING WIRES

THIN TUNGSTEN WIRE FILAMENT

FUSE

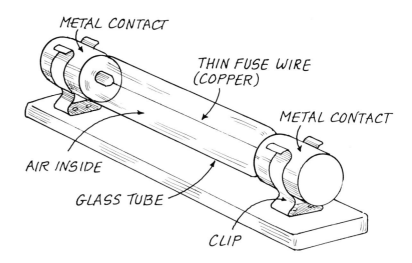

METAL CONTACT

THIN FUSE WIRE (COPPER)

METAL CONTACT

AIR INSIDE

GLASS TUBE

CLIP

Fuse wire gets very hot when the current through it is too big. It melts and breaks the circuit. Fuses are used to protect equipment from being damaged by currents which are too large.

REVISION QUESTIONS

1 When a fuse melts, does it open the circuit or does it close the circuit?

2 Why are electricity wires usually made from copper?

3 From which metal is the filament of a lamp made?

4 Why does an electrician's screwdriver have a plastic handle?

5 Which part of the breadboard is made from a conductor?

6 What can you say about all the metals you have tested?

7 Why is it dangerous to have an electric power-point in a bathroom?

8 Name a solid substance which is not a metal, but is a conductor.

9 A current is 1200 mA. What is this, in amps?

10 A current is 2.45 A. What is this, in milliamps?

11 A current is 850 mA. What is this, in amps?

8 Resistance

Your tests have shown you that some conductors are better than others at carrying current. You could list them in order, like this:

metals   *good conductors*
carbon   *not as good as metals*
salt water   *a poor conductor*
water   *a worse conductor*

Resistance is measured in ohms. *The short way of writing 'ohm' is to use the Greek letter 'omega':*

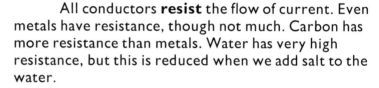

All conductors **resist** the flow of current. Even metals have resistance, though not much. Carbon has more resistance than metals. Water has very high resistance, but this is reduced when we add salt to the water.

The higher the resistance, the smaller the current.

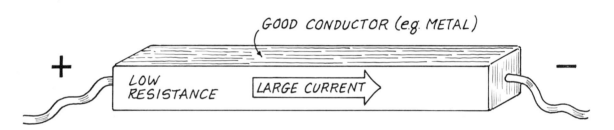

GOOD CONDUCTOR (e.g. METAL)

+   −

LOW RESISTANCE   LARGE CURRENT

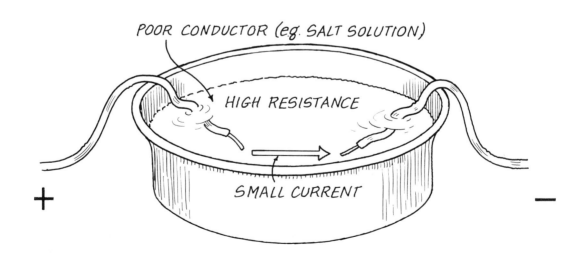

POOR CONDUCTOR (e.g. SALT SOLUTION)

HIGH RESISTANCE

+   −

SMALL CURRENT

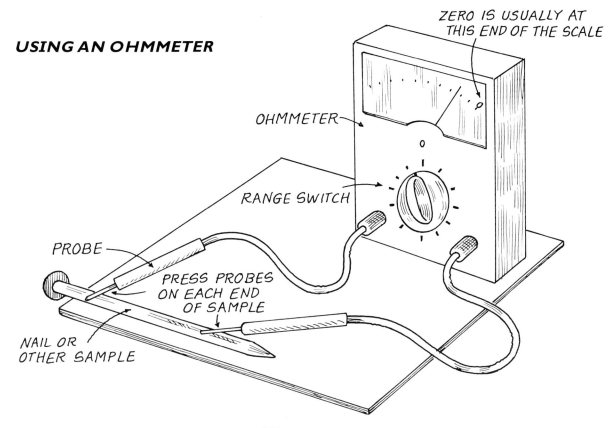

USING AN OHMMETER

ZERO IS USUALLY AT THIS END OF THE SCALE

OHMMETER

RANGE SWITCH

PROBE

PRESS PROBES ON EACH END OF SAMPLE

NAIL OR OTHER SAMPLE

We can measure resistance by using an **ohmmeter**. This tells us the resistance of a piece of conductor between its probes. Most test-meters can be switched to measure resistance, so you can use a test-meter as an ohmmeter.

Use an ohmmeter (or test-meter) to measure the resistance of pieces of conductor. Try these:

> a nail
> a short piece of thick copper wire
> carbon (pencil 'lead')
> tap water
> salt water
> coils of an electric motor, an electric bell and a loudspeaker; these are copper wires which are very long
> the filament of a lamp; this is a short wire but it is very thin
> your finger

Some of these have high resistance. You will need to alter the range switch on the meter for these. Write their resistances in **kilohms** or **megohms**, as explained on page 19.

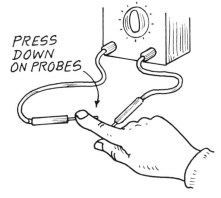

PRESS DOWN ON PROBES

This is how to measure the resistance of your finger.

9 Resistors

WIRE WOUND

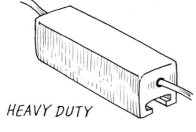

HEAVY DUTY

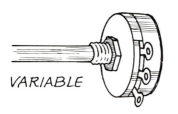

VARIABLE

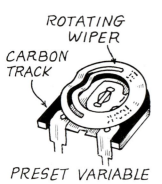

ROTATING WIPER

CARBON TRACK

PRESET VARIABLE

In electronics, we can use resistance to control the amount of current flowing round a circuit. We use **resistors** to give us the amount of resistance we want. We begin with some non-conducting material which is ground into a powder. Then a small amount of powdered carbon is mixed with this, so the mixture will be able to pass a small current. The mixture is moulded into small rods. Wires are fixed to each end.

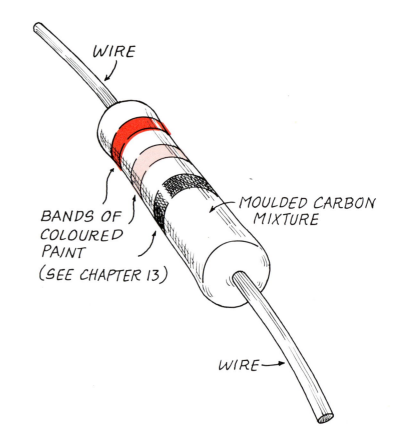

WIRE

BANDS OF COLOURED PAINT
(SEE CHAPTER 13)

MOULDED CARBON MIXTURE

WIRE →

The more non-conductor in the mixture, the higher the resistance of the rod.

The more carbon in the mixture, the lower the resistance of the rod.

In this way we can make resistors with the amount of resistance we want.

RESISTORS IN SERIES

Use the ohmmeter to measure the resistance of several different carbon resistors. Find two resistors which have resistances between 100 Ω and 1 kΩ. Measure the resistance of each carefully, and write down the results.

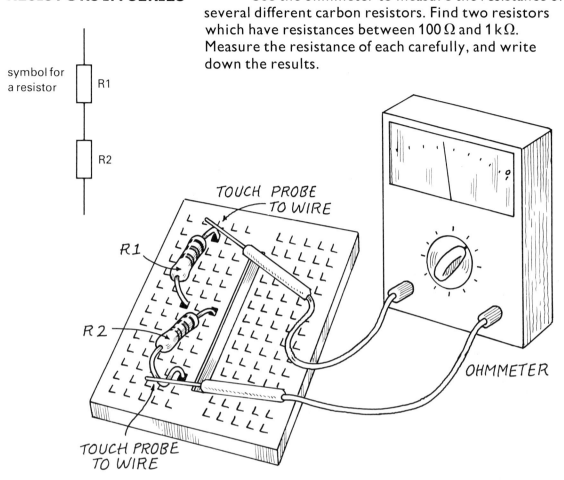

symbol for a resistor

R1

R2

TOUCH PROBE TO WIRE

R1

R2

TOUCH PROBE TO WIRE

OHMMETER

HIGH RESISTANCES

1000 ohms = 1 kilohm
The symbol for kilohm is kΩ.

1000 kilohms = 1 megohm
The symbol for megohm is MΩ.

In some books, resistors are drawn like this:

Now plug the resistors into a breadboard so that they are in series. Measure the resistance of the two resistors together. Write it down. What do you notice about their resistances separately and their resistance together, in series?

If you are not sure of your result, find two more resistors between 100 Ω and 1 kΩ and repeat the test.

10 Making currents flow

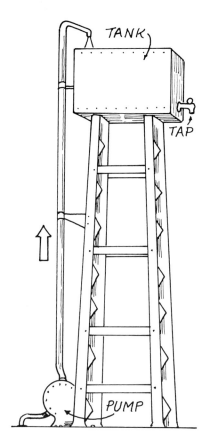

An electron (or other charge carrier) moves only if it is given energy. It gets its energy from a battery or power-pack. A battery is made of cells. Each cell contains chemicals. Energy is stored in the cell as **chemical energy**.

When the cell is in an electric circuit, its chemical energy is gradually turned into **electrical energy**. The electrical energy creates a force which moves electrons or other charge carriers. We could call this force the 'electric-charge moving force', or the **electromotive force**. 'Electromotive force' is often shortened to 'e.m.f.'

The power for a power-pack comes from a power-station. There, **mechanical energy** makes the generators turn. When the generators turn, mechanical energy is turned into electrical energy. This creates an e.m.f. The e.m.f. makes the current flow to our homes and factories, where it is used.

The e.m.f. of a battery (or generator) creates a **potential difference** (p.d.) between its terminals. One terminal is positive and the other is negative. The p.d. makes current flow from the positive terminal to the negative terminal.

FORCE CREATES POTENTIAL

The pump FORCES the water up to the high tank. It gives POTENTIAL ENERGY to the water. The water stays in the tank until the tap is opened.

When the tap is opened the water runs out. It falls to the ground. Its potential energy, given to it by the force of the pump, makes it MOVE. Its potential energy is turned into energy of movement (KINETIC ENERGY).

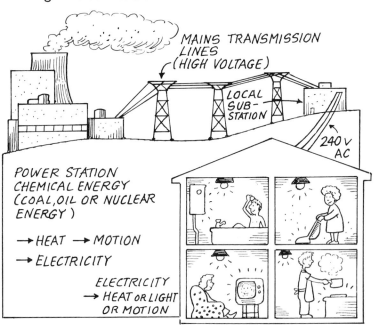

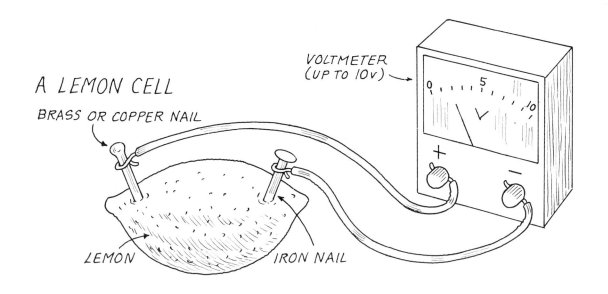

A LEMON CELL

BRASS OR COPPER NAIL

VOLTMETER (UP TO 10v)

LEMON

IRON NAIL

Use a voltmeter to measure the p.d.s of

a 'lemon' cell	a battery of two dry cells
a simple wet cell	a battery of three dry cells
a dry cell	a battery of four dry cells

We measure p.d. in volts, symbol V.

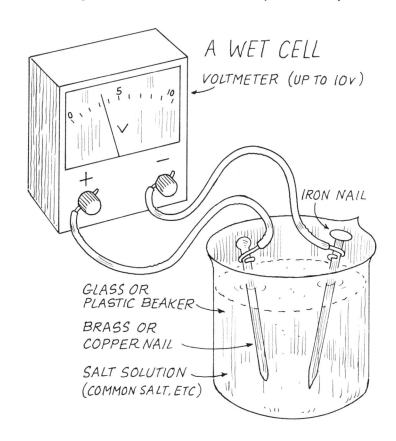

A WET CELL

VOLTMETER (UP TO 10v)

IRON NAIL

GLASS OR PLASTIC BEAKER

BRASS OR COPPER NAIL

SALT SOLUTION (COMMON SALT, ETC)

LIBRARY

262750·7

21

11 Potential differences

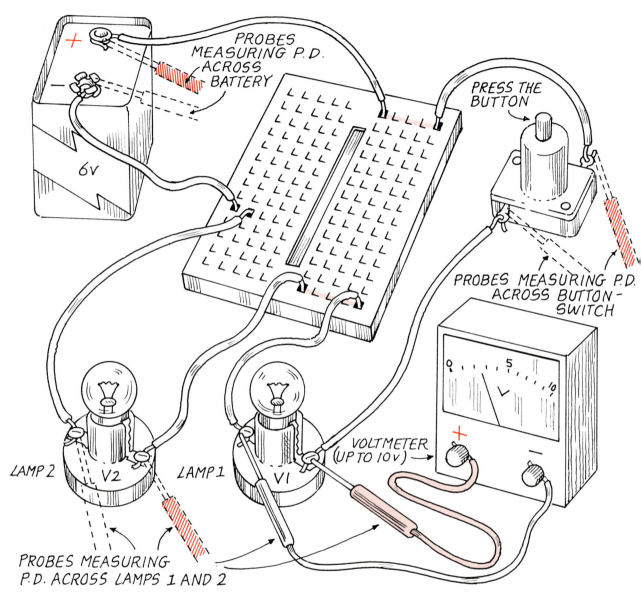

PROBES MEASURING P.D. ACROSS BATTERY

+

−

6V

PRESS THE BUTTON

PROBES MEASURING P.D. ACROSS BUTTON-SWITCH

VOLTMETER (UP TO 10V)

0 . . . . . 5 10

+

−

LAMP 2   V2   LAMP 1   V1

PROBES MEASURING P.D. ACROSS LAMPS 1 AND 2

Wire a circuit with two lamps in series (as on page 4). Measure the p.d. across the battery (which has four dry cells) – call this V. Measure the p.d.s. across each of the lamps. Call these V_1 and V_2.

Add $V_1 + V_2$. What does this sum equal?

Measure the p.d. across the button-switch, when it is not pressed and when it is pressed. Why does the p.d. change?

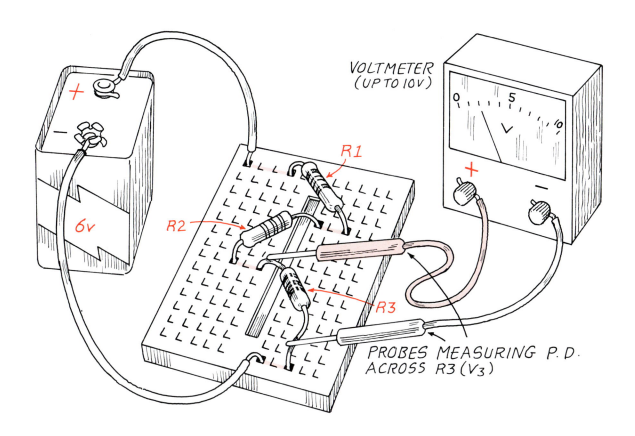

VOLTMETER
(UP TO 10V)

R1

R2

R3

PROBES MEASURING P.D.
ACROSS R3 (V₃)

6V

FORCE CREATES POTENTIAL

The E.M.F. of a cell gives the electrons POTENTIAL ENERGY. When the button is pressed the potential energy makes the electrons MOVE. Their potential energy is turned into KINETIC ENERGY.

On the way round the circuit they lose some of this energy, because of the resistance of the circuit. The energy may be turned into HEAT and LIGHT.

Use an ohmmeter to find three different resistors between 100 Ω and 1 kΩ. Measure the resistance of each. Call them R_1, R_2 and R_3, and make a note of them. Measure the p.d. of the battery, and call this V. Measure the p.d.s across each of the three resistors, and call them V_1, V_2 and V_3.

What does $V_1 + V_2 + V_3$ equal?
Which resistor has the biggest resistance?
Which resistor has the biggest p.d.?
Which resistor has the smallest resistance?
Which resistor has the smallest p.d.?

The bigger the resistance, the bigger the p.d. across it. Is the p.d. proportional to the resistance? Let us check this. Work out these fractions:

V_1 divided by R_1 V_2 divided by R_2
V_3 divided by R_3

What do you notice about these results?

12 Volts, amps and ohms

If you add an ammeter to the circuit on page 23, you can find out how volts, amps and ohms are related. Use resistors of the values shown.

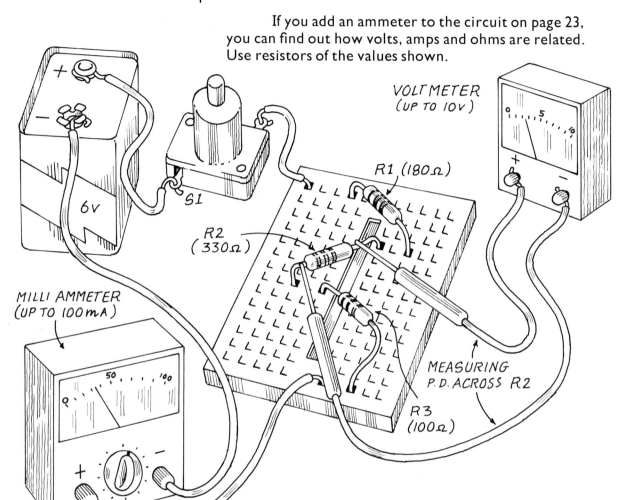

VOLTMETER
(UP TO 10V)

R1 (180Ω)

R2
(330Ω)

S1

6V

MILLI AMMETER
(UP TO 100mA)

MEASURING
P.D. ACROSS R2

R3
(100Ω)

Press the button and read the ammeter. How much current is flowing through it? How much current do you think is flowing through R1? How much through R2? How much through R3?

The battery, R1, R2, R3, and the ammeter are all in series, so the same amount of current flows through each. It is about 10 mA, or 0.01 A.

Press the button and use the voltmeter to measure the p.d. across R1. Divide the p.d. by the resistance. Do the same for R2 and for R3. What do you notice about the three answers?

The three answers are almost the same and they all have the value of the current, in amps.

The symbol for an ammeter is

The symbol for a voltmeter is

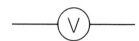

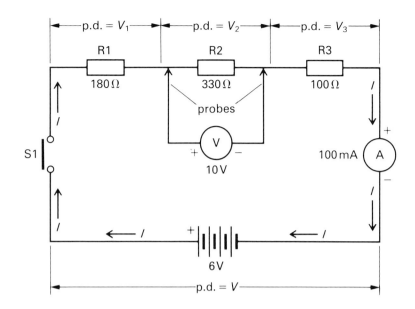

As a check, repeat this using three different resistors. Change R1 to 56 Ω, R2 to 220 Ω and R3 to 82 Ω. Does V divided by R come to the same value for all three resistors? Does this value equal the current, in amps? The current should be about 17 mA, or 0.17 A.

Try again, with any three of the resistors, but use a 3 V battery. Is the rule still true?

The VOLTS, AMPS and OHMS rule

Your results lead you to a rule. We use this rule a lot in electronics.

If we have a resistance R ohms, with a p.d. of V volts across it, and the current flowing through it is I amps, then:

$$I = \frac{V}{R}$$

This is the rule which relates volts, amps and ohms.

13 More about resistors

Resistors are usually marked with four coloured bands, like this:

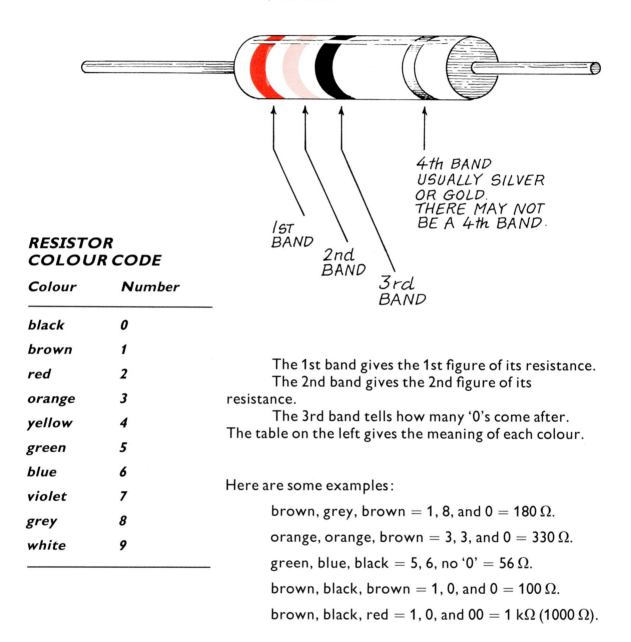

1ST BAND

2nd BAND

3rd BAND

4th BAND USUALLY SILVER OR GOLD. THERE MAY NOT BE A 4th BAND.

RESISTOR COLOUR CODE

Colour	Number
black	0
brown	1
red	2
orange	3
yellow	4
green	5
blue	6
violet	7
grey	8
white	9

The 1st band gives the 1st figure of its resistance.
The 2nd band gives the 2nd figure of its resistance.
The 3rd band tells how many '0's come after.
The table on the left gives the meaning of each colour.

Here are some examples:

brown, grey, brown = 1, 8, and 0 = 180 Ω.

orange, orange, brown = 3, 3, and 0 = 330 Ω.

green, blue, black = 5, 6, no '0' = 56 Ω.

brown, black, brown = 1, 0, and 0 = 100 Ω.

brown, black, red = 1, 0, and 00 = 1 kΩ (1000 Ω).

TEST YOURSELF You need ten assorted resistors. Use the colour code to find out their resistance. Then measure their resistance with an ohmmeter to see if you are right.

SUMMING UP So far, you have found out the following:

1 The same current (I) flows through each resistor.
2 There is a p.d. (V) across each resistor.
3 $I = V/R$, for each resistor.
4 The total of the p.d.s across each resistor is the same as the p.d. across the battery.
5 Total resistance $R = R_1 + R_2 + R_3$ for resistors in series.

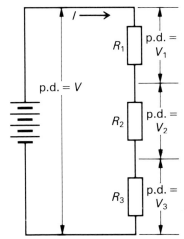

REVISION QUESTIONS

1 *In what unit do we measure resistance?*

2 *You have two resistors, 120 Ω and 390 Ω. What is their total resistance when they are in series?*

3 *What do we call the electrical force which is generated by the chemical energy stored in a cell?*

4 *What instrument is used for measuring (a) current, (b) resistance, and (c) p.d.?*

5 *What are the resistances of resistors marked (a) red, red, red; (b) brown, black, orange; (c) yellow, violet, yellow?*

6 *What are the colours of the bands on the following resistors: (a) 680 Ω, (b) 680 kΩ, (c) 1 MΩ, (d) 33 Ω?*

7 *What is the volts-amps-ohms rule?*

8 *If there is a p.d. of 10 V across a resistance of 5 Ω, what current flows through the resistance?*

9 *Two resistors, 10 kΩ and 15 kΩ, are in series. There is a p.d. of 25 V across the pair. What is the p.d. across each resistor?*

14 Resistors in parallel

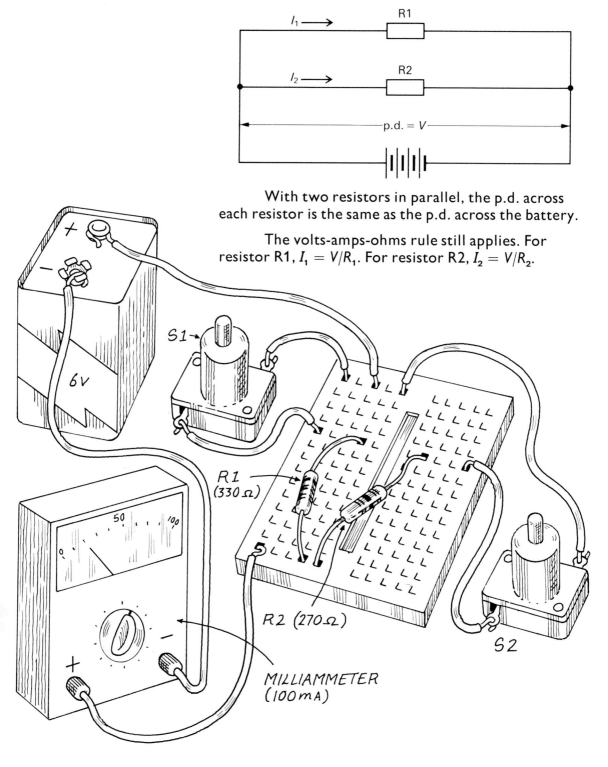

With two resistors in parallel, the p.d. across each resistor is the same as the p.d. across the battery.

The volts-amps-ohms rule still applies. For resistor R1, $I_1 = V/R_1$. For resistor R2, $I_2 = V/R_2$.

Press button 1 and measure the current through R1.

Press button 2 and measure the current through R2.

Press both buttons at once and measure how much current can pass through both resistors.

The total current is $I = I_1 + I_2 = V/R_1 + V/R_2$.

Check this by working out I_1 and I_2, and their total, I.

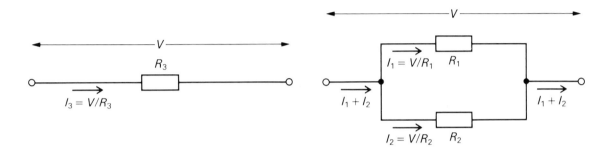

We could put a single resistor R3 in place of R1 and R2 to pass the same current. We will work out what resistor to put in. To begin with we know that

and
$$I_3 = V/R_3$$
$$I_3 = I_1 + I_2 = V/R_1 + V/R_2.$$

So
$$\frac{V}{R_3} = \frac{V}{R_1} + \frac{V}{R_2}$$

Dividing both sides of the equation by V gives

$$\frac{1}{R_3} = \frac{1}{R_1} + \frac{1}{R_2}$$

If you work out this for the two resistors,

$$1/R_3 = 1/330 + 1/270 = 0.00303 + 0.00370 = 0.00673$$

so $R_3 = 1/0.00673 = 149\ \Omega$.

The replacement resistor, R3, should have a resistance 149 Ω. Take R1 and R2 away and put a 150 Ω resistor in place of R1. Press button 1. Is the current 40 mA, the same as for R1 and R2 in parallel?

15 Summing up 1

These are the main points we have covered so far.

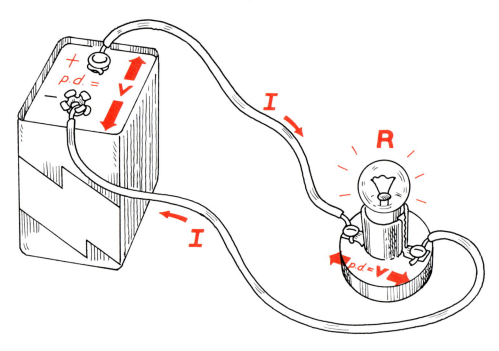

UNITS   **Current** is measured in **amps**, using an **ammeter**.

Potential difference is measured in **volts**, using a **voltmeter**.

Resistance is measured in **ohms**, using an **ohmmeter**.

SYMBOLS   The symbol of amps is A. A **milliamp** (mA) is a thousandth of an amp. A **microamp** (μA) is a millionth of an amp.

The symbol for volts is V. A **millivolt** (mV) is a thousandth of a volt. A **kilovolt** (kV) is a thousand volts.

The symbol for ohms is Ω. A **kilohm** (kΩ) is a thousand ohms. A **megohm** (MΩ) is a million ohms.

VOLTS-AMPS-OHMS RULE   In any circuit or part of a circuit,
$$V = IR \quad \text{or} \quad I = \frac{V}{R} \quad \text{or} \quad R = \frac{V}{I}.$$

This rule applies to anything which has resistance, such as a piece of wire, a lamp, or a resistor.

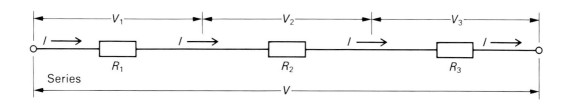

Series

RESISTANCES IN SERIES

The same current passes through each resistor. The total resistance is $R_1 + R_2 + R_3 \ldots$ (simply add the resistances together).

The total p.d. (V) across all resistors can be split into $V_1, V_2, V_3, \ldots$

The volts-amps-ohms rule applies to each resistor, so

$$V_1 = IR_1, \quad V_2 = IR_2, \quad V_3 = IR_3 \ldots$$

and $\quad V = V_1 + V_2 + V_3 \ldots$

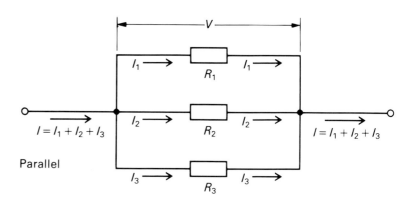

Parallel

RESISTANCES IN PARALLEL

The p.d. is the same across all resistors.

The total resistance R is found from:

$$\frac{1}{R} = \frac{1}{R_1} + \frac{1}{R_2} + \frac{1}{R_3} + \ldots$$

The volts-amps-ohms rule applies to each resistor, so $\quad I_1 = V/R_1, \quad I_2 = V/R_2, \quad I_3 = V/R_3, \ldots$ The current through each resisistor is inversely proportional to its resistance (e.g. if one resistor has twice the resistance of another, it passes half as much current).

$$I = I_1 + I_2 + I_3 + \ldots$$

16 Dividing potentials

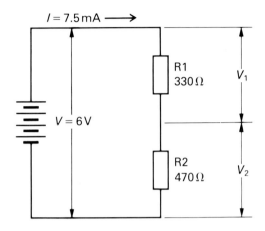

When we have two resistors in series, the same current passes through both of them. In the example,

$$I = \frac{V}{R_1 + R_2} = \frac{6}{330 + 470} \qquad (R_1 + R_2 = R)$$

$$= \frac{6}{800} = 0.0075 \text{ A}.$$

We have used the volts-amps-ohms rule to find the current.

When the current flows, there is a p.d. across each resistor. The p.d.s must add up to the p.d. of the battery (6 V). We work out each p.d. by using the form of the volts-amps-ohms rule which says

$$V = I \times R \qquad \text{(see page 31)}.$$

The p.d.s across the two resistors are

$$V_1 = I \times R_1 = 0.0075 \times 330 = 2.475 \text{ V}$$
$$V_2 = I \times R_2 = 0.0075 \times 470 = 3.525 \text{ V}$$

Check the working by adding the two p.d.s: $2.475 + 3.525 = 6$ V, as expected.

This is a useful circuit. If you need a p.d. of 3.5 V but you only have a 6 V battery, the potential divider gives you the 3.5 V you need.

The circuit divides the p.d. of the battery (6 V) into two parts (2.475 V and 3.525 V). This kind of circuit is called a **potential divider**.

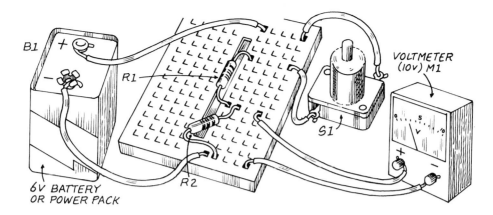

6V BATTERY
OR POWER PACK

R1

R2

VOLTMETER
(10v) M1

S1

B1

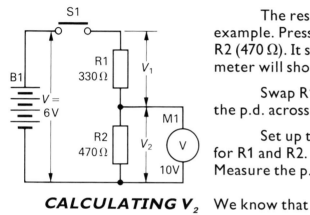

S1

B1

$V = $
6V

R1
330 Ω

V_1

R2
470 Ω

V_2

M1

V

10V

CALCULATING V_2

The resistors in this circuit are the same as the example. Press the button and measure the p.d. across R2 (470 Ω). It should be 3.525 V, but the nearest the meter will show is about 3.5 V. Do you get this?

Swap R1 with R2 and measure again. Measure the p.d. across R1 (330 Ω). It should be about 2.5 V.

Set up the circuit again with different resistors for R1 and R2. R1 can be 1 kΩ and R2 can be 560 Ω. Measure the p.d. across R2.

We know that
$$I = \frac{V}{R_1 + R_2}$$

Putting this value for I in the next equation, we get

$$V_2 = I \times R_2 = \frac{V}{R_1 + R_2} \times R_2 = V \times \frac{R_2}{R_1 + R_2}.$$

If R1 is 1 kΩ, R2 is 560 Ω and $V = 6$ V, then

$$V_2 = V \times \frac{R_2}{R_1 + R_2} = 6 \times \frac{560}{1560} = 2.15 \text{ V}.$$

Did you find this when you measured V_2?

Try using other pairs of resistors. Work out what V_2 should be. Then use the meter to find out if you are right.

You are given a 6 V battery and four resistors with values 120 Ω, 120 Ω, 220 Ω and 680 Ω. Make drawings to show which resistors you would connect to the battery to make potential dividers giving

(a) 3 V, (b) 0.9 V, (c) 5.1 V, (d) 2 V.

Check your calculations by building each circuit and measuring the p.d. produced.

By choosing the right pair of resistors, you can make the divider give you any p.d. you need between 0 V and the full p.d. of the battery.

17 Electrons and heat

This circuit is a potential divider, like the circuit in chapter 16. But one of the resistors (R2) is of a special kind. It is a **thermistor**.

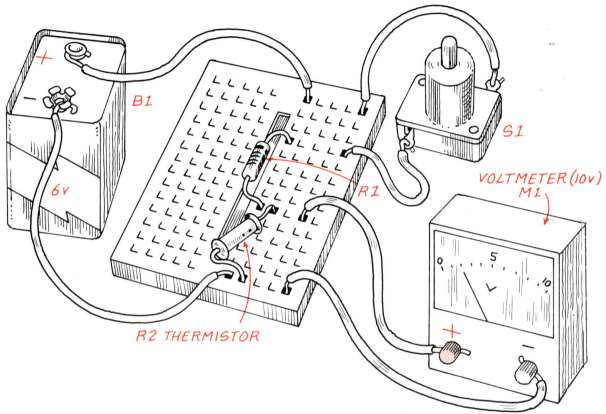

Press the button and read the p.d. across the thermistor (R2). Now grip the thermistor closely with your hand (but do not pull it out of the breadboard). Watch the needle of the meter. The p.d. is slowly falling. This **could** mean that the p.d. of the battery is slowly falling, but you can measure it, to check that this is not happening. The p.d. across the thermistor falls because **its resistance has become smaller**.

Take your hand away from the thermistor. Leave it a minute or two to cool down again. Press the button. Is the p.d. the same as when you began? If so, the resistance of the thermistor has gone back to what it was at the beginning.

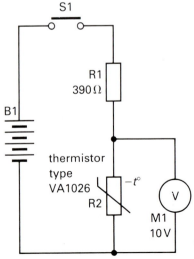

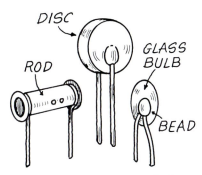

Thermistors are made in many different shapes and sizes.

Take a piece of ice and hold this against the thermistor. Press the button. What happens to the p.d.? What must be happening to the resistance of the thermistor?

The resistance of a thermistor changes with temperature.

When temperature goes up, its resistance goes down.

When temperature goes down, its resistance goes up.

In this way a thermistor is a **temperature-dependent resistor**.

The material a thermistor is made of is a poor conductor. It does not have many electrons to carry charge. When it gets warmer, some of the electrons can escape more easily from their atoms. This gives more charge carriers. Bigger currents can flow. The resistance of the material goes down. The opposite happens when the thermistor is made cold.

REVISION QUESTIONS

1   You have a 6 V battery and a 1 kΩ resistor. What other resistor do you need to make a potential divider which gives 4 V?

2   If, in the circuit opposite, the positions of R1 and the thermistor were swapped, what would happen to the needle of the meter when you put the thermistor near a warm fire?

A PROJECT

The reading on the meter depends on temperature. The scale of the meter could be marked to show temperatures instead of volts. It can be used as an electronic thermometer. Try making one. Mark its scale in degrees Celsius (°C).

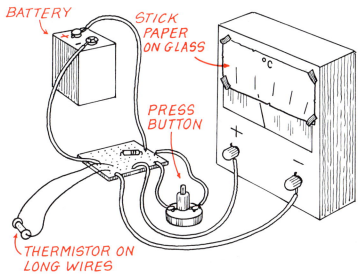

18 Semiconductors

Semiconductors are a group of materials which have small amounts of charge carriers in them. They conduct, but not as well as metals.

Cold semiconductors have very few charge carriers. Their resistance is high. When semiconductors are warmed, electrons escape from their atoms and become charge carriers. The warmer the semiconductor, the more electrons escape and the better it conducts. The warmer it is, the lower its resistance.

Thermistors (chapter 17) are made from a semiconductor. This is why their resistance becomes less as they become warmer.

The most commonly-used semiconductor is silicon. Small amounts of other substances can be added to it, to give it more charge carriers. Different substances give two different types of silicon. One is called n-type silicon and the other is called p-type silicon.

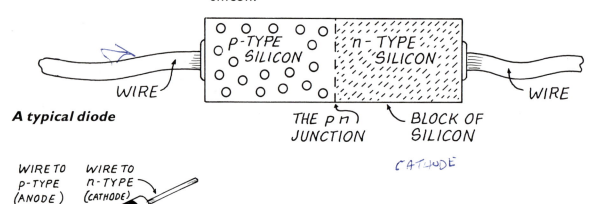

A typical diode

WIRE

P-TYPE SILICON

n - TYPE SILICON

THE p n JUNCTION

BLOCK OF SILICON

WIRE

CATHODE

WIRE TO P-TYPE (ANODE)

WIRE TO n-TYPE (CATHODE)

PAINTED BAND ON CASE

We use a diode when we want to let a current flow in one direction but not in the other (more about this in chapter 43).

To make a **diode** we take a piece of silicon, make one part of it into n-type silicon and the other part into p-type silicon.

Where the two parts join is called the pn junction. This junction gives the diode its special feature.

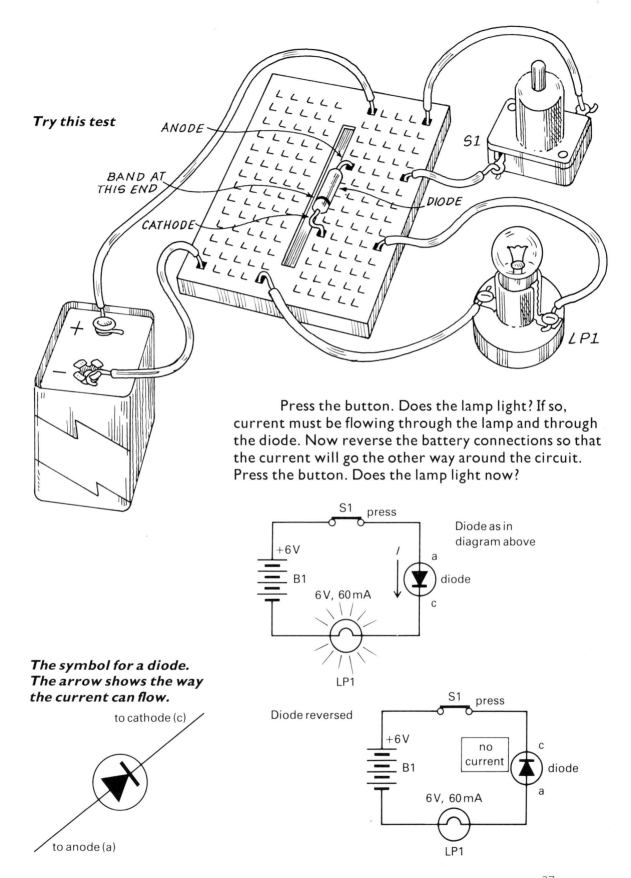

Try this test

ANODE

BAND AT
THIS END

CATHODE

S1

DIODE

LP1

Press the button. Does the lamp light? If so, current must be flowing through the lamp and through the diode. Now reverse the battery connections so that the current will go the other way around the circuit. Press the button. Does the lamp light now?

S1 press

+6V

B1

6V, 60mA

Diode as in diagram above

a

diode

c

LP1

The symbol for a diode. The arrow shows the way the current can flow.

to cathode (c)

to anode (a)

Diode reversed

S1 press

+6V

B1

no current

c

diode

a

6V, 60mA

LP1

19 More about diodes

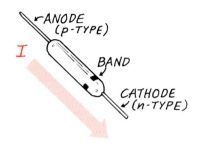

ANODE (p-TYPE)

BAND

CATHODE (n-TYPE)

The tests in chapter **18** showed that electric current can easily flow from the p-type to the n-type silicon. It **cannot** flow from the n-type to the p-type silicon. **Diodes conduct in one direction only**.

The connecting wires of a diode are sometimes called **anode** (to the p-type end) and **cathode** (to the n-type end). Current flows in at the anode and out of the cathode.

Some diodes are made from semiconductors of a special kind. When a current flows through these diodes, the atoms at the pn junction give off light. These diodes are called **light-emitting diodes**, or LEDs for short. Try this circuit. Do not forget to include R1. It makes the current small, so it will not damage the light-emitting diode.

A typical light-emitting diode (LED)

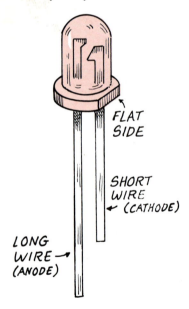

FLAT SIDE

SHORT WIRE (CATHODE)

LONG WIRE (ANODE)

The symbol for an LED

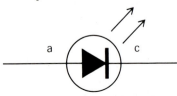

a   c

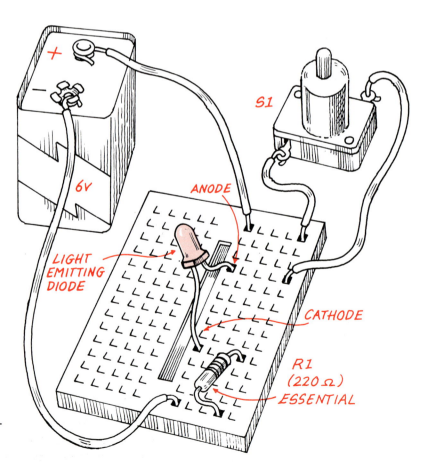

S1

6v

ANODE

LIGHT EMITTING DIODE

CATHODE

R1 (220 Ω) ESSENTIAL

Reverse the connections of the diode. Does the LED light now?

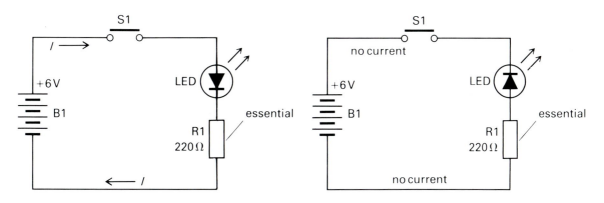

The above diagrams show what happens when switch S1 is pressed.

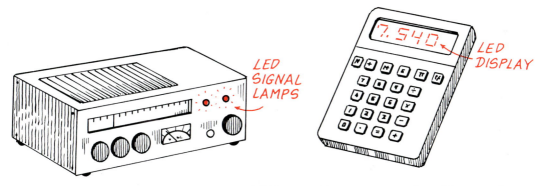

LED SIGNAL LAMPS

LED DISPLAY

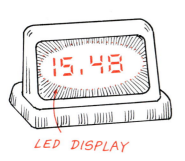

LED DISPLAY

LEDs need less current than filament lamps:

	LED	Small filament lamp (torch bulb)
To glow dimly	0.25 mA	20 mA
To shine brightly	20 mA	60 mA

LEDs are useful as signal lamps (e.g. 'power on') or as LED displays. They are specially useful in battery-powered equipment, to save current. Also, they last much longer than filament lamps.

LEDs cannot be made bright enough for other jobs which filament lamps do. They are not bright enough for lighting rooms or electric torches.

20 Transistors

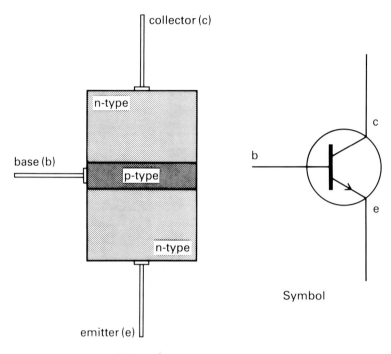

collector (c)

base (b)

p-type

n-type

n-type

emitter (e)

Diagram

c

b

e

Symbol

COLLECTOR

BASE
EMITTER

Like a diode, a **transistor** is made from a piece of semiconducting material, usually silicon. A diode has only two regions, p-type and n-type, but a transistor has three regions. Most transistors have a thin region of p-type silicon sandwiched between two layers of n-type silicon. These transistors are called **npn transistors**. The two n-type regions are called the **collector** (c) and the **emitter** (e). The p-type region is called the **base** (b).

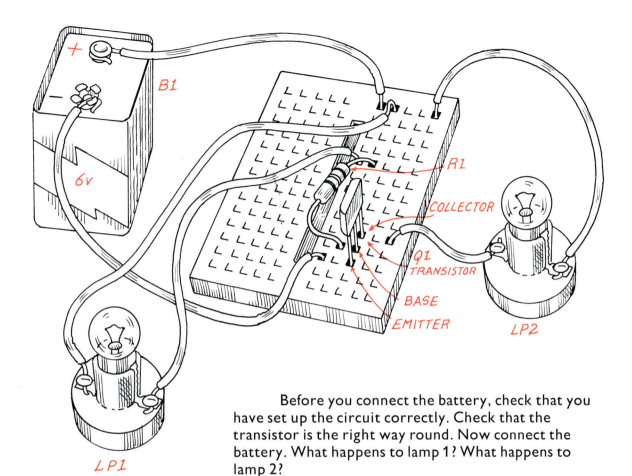

Before you connect the battery, check that you have set up the circuit correctly. Check that the transistor is the right way round. Now connect the battery. What happens to lamp 1? What happens to lamp 2?

We can see that a current is flowing through lamp 2, because it is shining brightly. The current is about 60 mA. Lamp 1 is not even glowing, so the current through it must be small. Maybe there is no current. Take lamp 1 out of its socket. What happens to lamp 2?

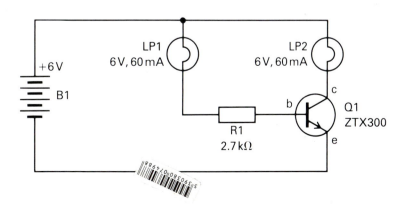

21 Transistor switches

This is how the currents flow.

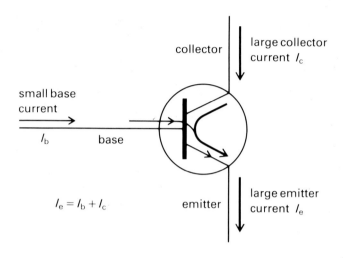

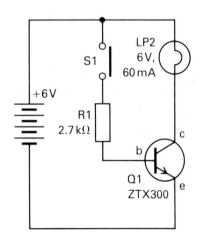

Set up the same circuit as in chapter 20, with a button-switch in the place of lamp 1. Press the button. This lets a small current flow to the base. We can work out how small this is.

If we ignore the small p.d. between the base and emitter of the transistor, the p.d. across R1 is 6 V. Now we can use the volts-amps-ohms rule (page 31):

$$V = 6 \text{ V}, R = 2700 \text{ }\Omega, I = 6/2700 = 0.002 \text{ A or 2 mA.}$$

We switch a current of 2 mA on or off, making the transistor switch a current of about 60 mA on or off.

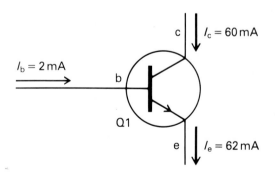

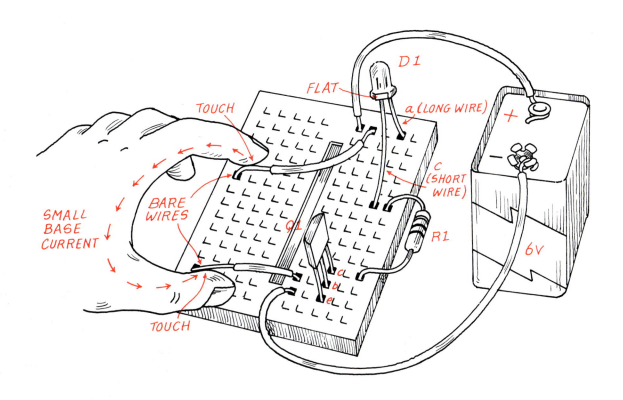

Wet your thumb and first finger. Touch your thumb to one wire and your finger to the other. Watch the LED. When you touch the wires, a very small current can flow through your hand from one wire to the other, and then to the base of the transistor. The transistor switches on a much larger current to light the LED.

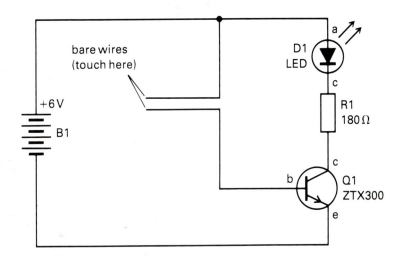

22 Using transistors

**CROSSING
THE PN JUNCTIONS**

On page 38 it says that current can flow across a pn junction from p-type to n-type. It cannot flow from n-type to p-type. There are two pn junctions in a transistor. The current flowing into the base goes across the junction between the base and the emitter. The base current flows easily from p-type to n-type. The current flowing into the collector first has to cross the junction between collector and base. It has to flow from n-type to p-type. We have said that this is impossible, so how can it happen?

The base is a very thin layer. When the base current flows to the emitter, it also goes very close to the collector. This closeness helps the collector current to cross the junction. The collector current is then able to enter the base. From there, it can easily cross the next junction, from base to emitter.

If there is no base current, there can be no collector current, for there is nothing to help it across the junction.

A WATER-LEVEL DETECTOR

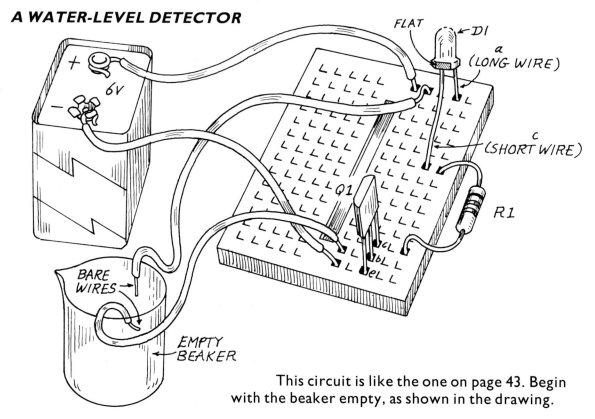

This circuit is like the one on page 43. Begin with the beaker empty, as shown in the drawing.

There is a gap between the wires in the beaker. No current can flow to the base of the transistor. No base current means no collector current. The LED is off.

Now slowly pour water into the beaker. What happens when the water reaches the wires? Water is not a good conductor. It has very high resistance. The current going to the base is very small. Yet it is enough to help a large collector current cross the pn junction. The large collector current goes through the LED and makes it shine. This circuit could be used as a 'water-tank overflowing' warning.

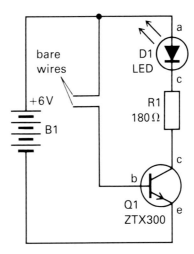

REVISION QUESTIONS

1   In which direction can current flow through a diode?

2   Name the three regions of an npn transistor. Which region is made from p-type material?

3   In what way does the resistance of a semiconductor change when it is warmed?

4   Which is the most commonly used semiconducting material?

5   Where is the light generated in a light-emitting diode?

6   If you want to have a small lamp on a battery-powered radio to show that the power is switched on, what are the advantages of using a light-emitting diode, rather than a small filament lamp?

23 Switching by temperature

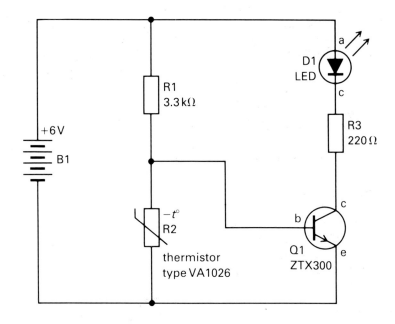

Can you spot a potential divider in this circuit? If not, look again at page 32. Look at page 34 as well, to remind yourself about thermistors.

To begin with, the resistance of the thermistor is high. The p.d. across it is high, as explained on page 34. This p.d. makes a base current flow, turning the transistor on. The collector current lights the LED.

Now warm the thermistor. Put it into a beaker of warm water or hold it close to a bench-lamp. Explain what happens (page 34 will help). This is another example of the transistor being used as a switch.

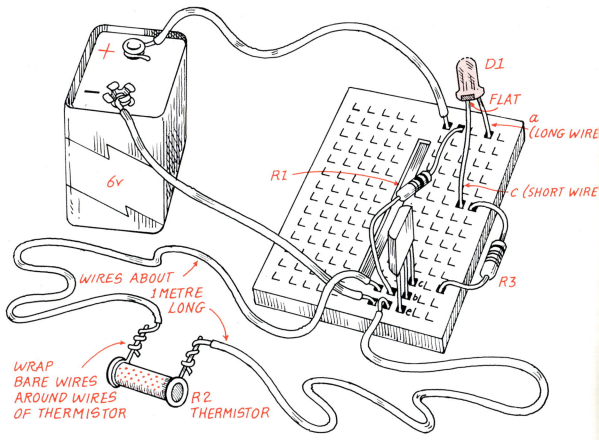

PROBLEM

Alter the circuit so that the LED is on at room temperature but goes off when the thermistor is cooled.

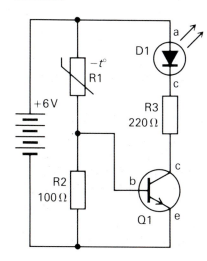

REVISION QUESTIONS

1   Which component in the diagram on the left is the thermistor?

2   At 25°C, the resistance of the thermistor is 1 kΩ. What is the p.d. across the thermistor at that temperature?

3   What will happen to the p.d. across the thermistor if we cool the thermistor?

4   The LED is on when the temperature of the thermistor is 25°C. What will happen to the LED when the thermistor is made very cold?

24 Electrons and light

A light-dependent resistor

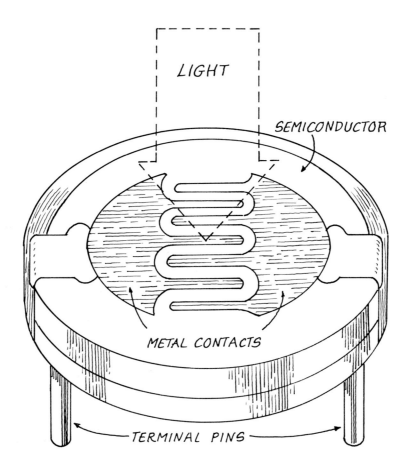

LIGHT

SEMICONDUCTOR

METAL CONTACTS

TERMINAL PINS

A **light-dependent resistor** (LDR) is a piece of semiconducting material. When light falls on it, electrons are set free from the atoms of the material. They can act as charge carriers. Now it is easier for a current to flow through the material from one contact to the other. The resistance of the LDR is made low. Few electrons are set free when the LDR is in the dark, so then its resistance is very high.

Another name for a light-dependent resistor (LDR) is photo-conductive cell (PCC).

Measure the resistance of an LDR using an ohmmeter or test-meter (page 17). See how its resistance changes when you cover the LDR with your hand. See how it changes when you shine a bright light on it.

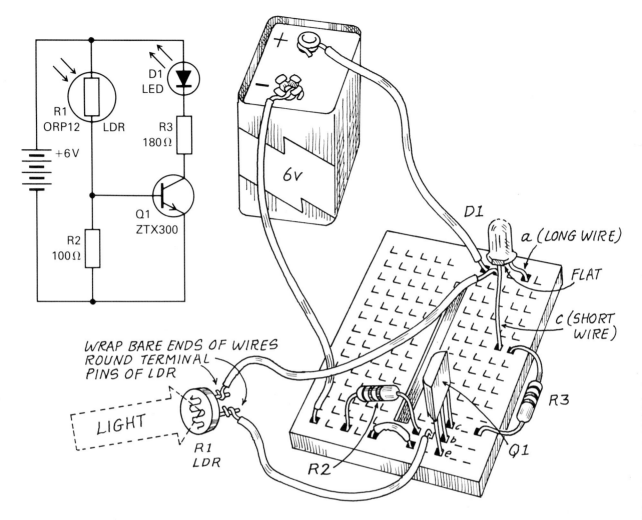

This circuit is like the one on page 47 in the revision questions. The resistor and the LDR make a potential divider. The transistor is switched on when a bright light shines on the LDR. Cover the LDR with your hand. What happens then? Can you explain how this circuit works?

REVISION QUESTIONS

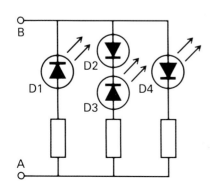

1   *Turn back to the diagram at the bottom left of page 47. When the transistor is switched on, a collector current of 20 mA flows through the LED (D1), R3 and the transistor (Q1). What is the p.d. across R3? If the p.d. across D1 is 1.6 V, what is the p.d. across Q1?*

2   *Which LEDs will light when terminal A is connected to the 0 V terminal of a battery and terminal B is connected to the + 6 V terminal of the battery? Which LEDs will light when the connections to the battery are reversed?*

25 Input and output

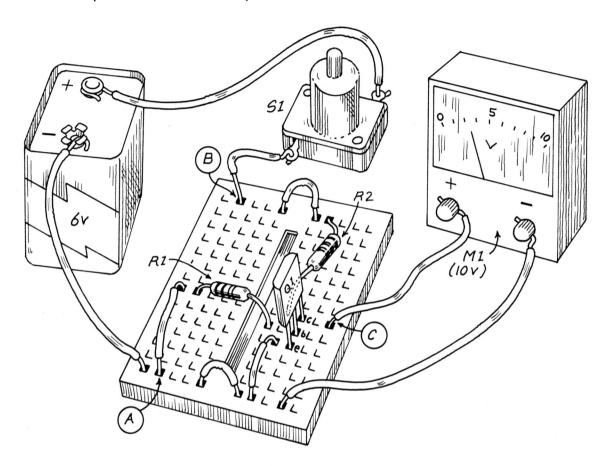

TRY THIS FIRST

Press the button to supply power to the circuit. Read the meter. The base of the transistor is at 0 V, because it is joined to the 0 V line. The voltage at the base is called the **input** to the transistor.

Because the input is 0 V, there is no base current. So there is no collector current. The transistor is switched off. No current is flowing through R2. This means that there is no p.d. across it:

$$R_2 = 100 \, \Omega, I = 0 \, A, \text{ so } V = I/R_2 = 0/100 = 0 \, V.$$

Because the p.d. across R1 is 0 V, the p.d. across Q1 must be 6 V. Therefore, the voltage at point C is +6 V. The voltage at point C (at the collector of the transistor) is called the **output** of the transistor.

When the input is 0 V, the output is +6 V.

The two p.d.s must add up to 6 V, which is the p.d. across the battery.

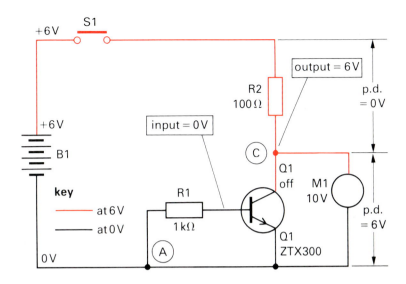

NOW TRY THIS ONE

The circuit below is like the one you have just tested. Take the end of the wire out of socket A and plug it into socket B. The base is now connected to the 6 V line. Press the button to supply power to the circuit. The input to the transistor is 6 V. Read the meter to find the output of the transistor.

When the input is 6 V, the output is 0 V.

The input is 6 V, so a base current flows. This makes a collector current flow. The transistor is switched fully on (we say it is **saturated**). It acts like a short-circuit between R1 and the 0 V line. The p.d across it is 0 V, so the p.d. across R1 must be 6 V. The voltage at the output is 0 V, as shown by the meter.

Once again, the p.d.s must add up to 6 V.

The arrows show the current which flows when S1 is pressed.

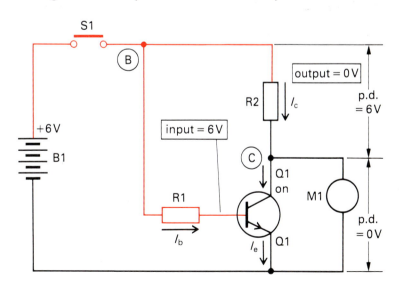

51

26 Rule of opposites

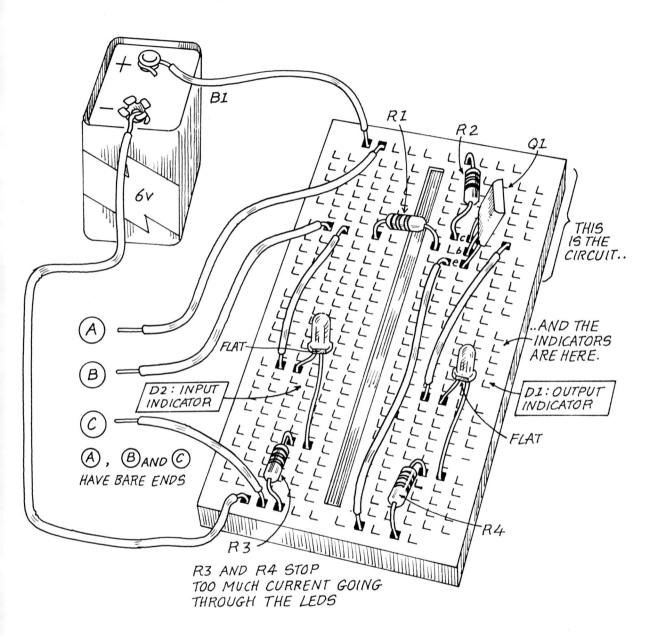

B1

6v

A

B

C

A, B AND C
HAVE BARE ENDS

FLAT

D2 : INPUT
INDICATOR

R1

R2

Q1

c
b
e

THIS
IS THE
CIRCUIT..

..AND THE
INDICATORS
ARE HERE.

D1 : OUTPUT
INDICATOR

FLAT

R4

R3

R3 AND R4 STOP
TOO MUCH CURRENT GOING
THROUGH THE LEDS

Here is another way to show input and output.
We use the LEDs as indicators. If the input or output is
0 V, the LED is off. If the input or output is 6 V, the LED
shines brightly. Touch the loose wire B first to wire A,
and then to wire C. When one LED is on, the other is
off.

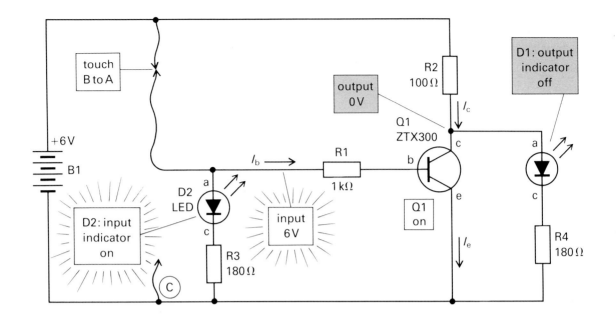

The rule for this circuit is a **rule of opposites**:

If the input is 0 V, the output is 6 V.
If the input is 6 V, the output is 0 V.

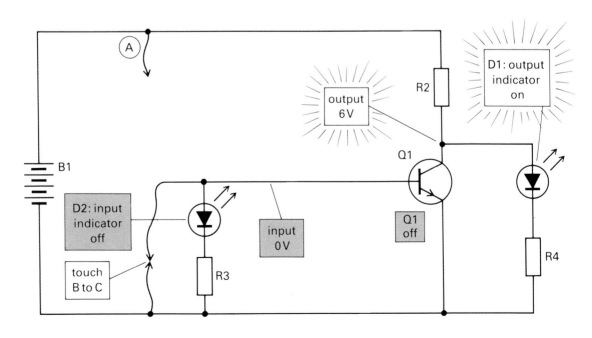

27 Output gives input

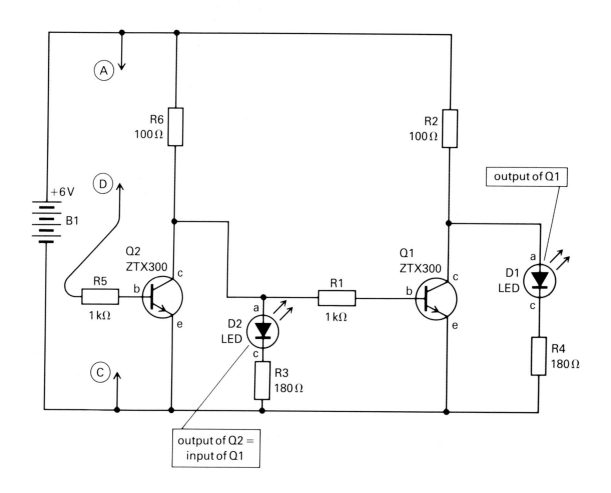

+6V
B1

A

D

C

R6
100 Ω

R2
100 Ω

R5
1 kΩ

Q2
ZTX300

R1
1 kΩ

Q1
ZTX300

D2
LED

R3
180 Ω

D1
LED

R4
180 Ω

output of Q1

output of Q2 =
input of Q1

Part of the circuit above is exactly the same as
the circuit on page 53. Can you see which part? But
instead of wire B there is the collector wire of a second
transistor (Q2). As a result of this, the **input** of Q1 is
always **the same** as the **output** of Q2. You change the
input of Q2 by touching wire D to wire A or wire C.
Try it.

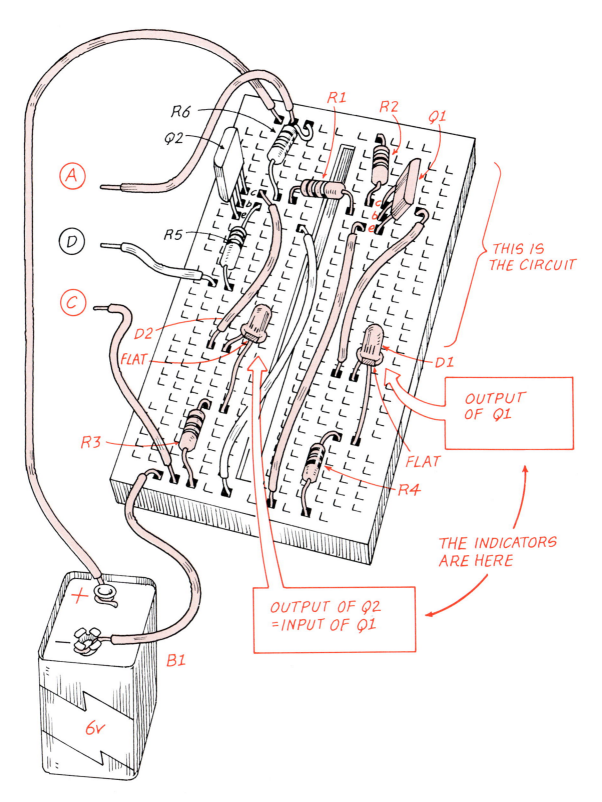

R6

R1

R2

Q1

Q2

A

D

R5

C

D2

FLAT

R3

D1

FLAT

R4

THIS IS
THE CIRCUIT

OUTPUT
OF Q1

THE INDICATORS
ARE HERE

OUTPUT OF Q2
=INPUT OF Q1

B1

6v

Keep this circuit wired up, for use in chapter 28.

28 A flip-flop

Use the circuit that you made for chapter 27. Take away wire A. Touch wire D to wire C. This makes Q2 turn off. Its output rises to 6 V. D2 comes on.

The 6 V output from Q2 turns Q1 on. Its output goes low. D1 goes out.

Wire D is at 0 V. The collector of Q1 is at 0 V. If you connect wire D to the collector of Q1 it should make no difference to the voltage of wire D. Try it. Plug wire D into the breadboard, as shown in the drawing below. You should find that D2 stays on and D1 stays off. There is no change.

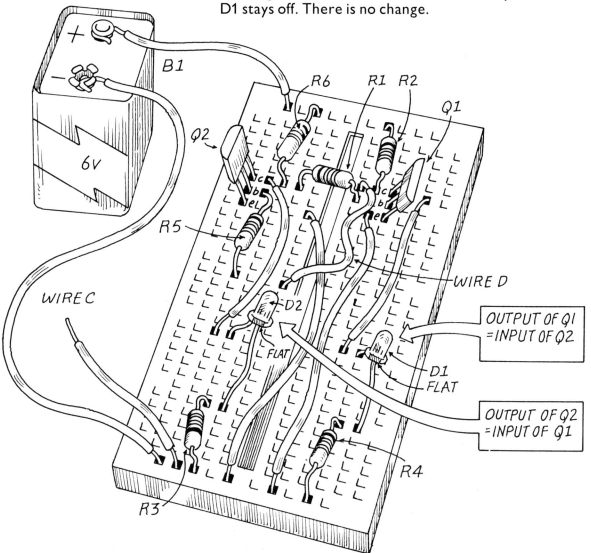

With wire D connected like this, the output of Q1 is the input of Q2. The output of Q2 is the input of Q1. The output of one transistor is the input of the other transistor. Each transistor is controlling its partner. The circuit cannot change state. It is **stable**.

You can make Q1 turn off by touching wire C against the wire of R1. Do this.

You have turned Q1 off. Its output goes to 6 V. D1 comes on. The 6 V output from Q1 turns Q2 on. Its output falls to 0 V. D2 goes out.

You can take wire C away now, because the 0 V output from Q2 keeps Q1 turned off. Q2 is holding Q1 off and Q1 is holding Q2 on. There is no further charge. The circuit is stable in this state too.

You have 'flipped' the circuit from one stable state to the other. This circuit is called a **flip-flop**.

Keep this circuit wired up for chapter 29.

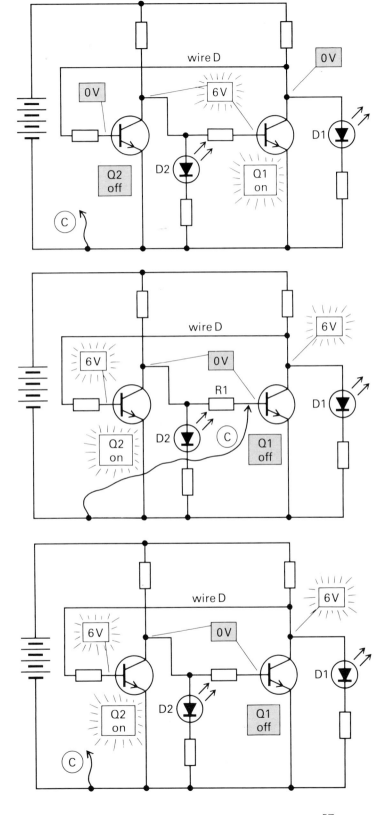

29 Two stable states

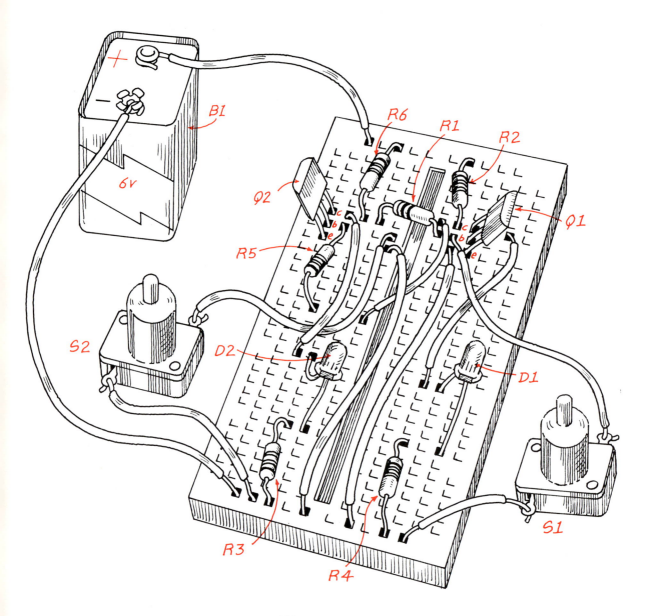

Wire two button-switches to the flip-flop circuit, as shown above. These make it easier for you to switch off the transistors in turn. Press S1 for an instant to switch off Q1, or press S2 for an instant to switch off Q2. Press each switch in turn. Work out what is happening in the circuit (remember the rule of opposites, chapter 26). Try pressing one switch over and over again. What happens to the circuit?

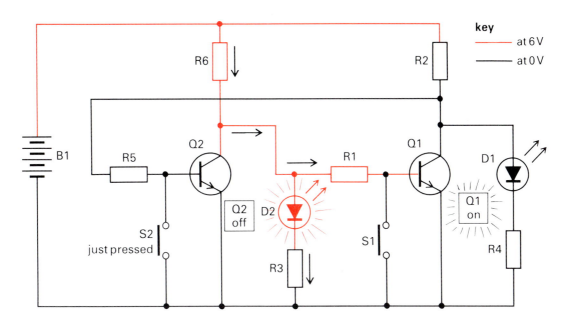

This is what the circuit is like in one of its stable states. It has just been 'flipped' into this state by pressing S1.

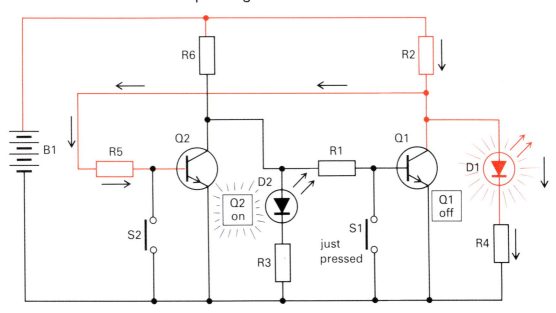

Now button S2 has just been pressed and the circuit has 'flopped' back to its other stable state. It has **two** stable states, so it is called a **bi**stable circuit.

The circuit can 'remember' which of the buttons was the last one to be pressed. Flip-flops similar to this are used in the memory circuits of calculators and computers.

30 Capacitors

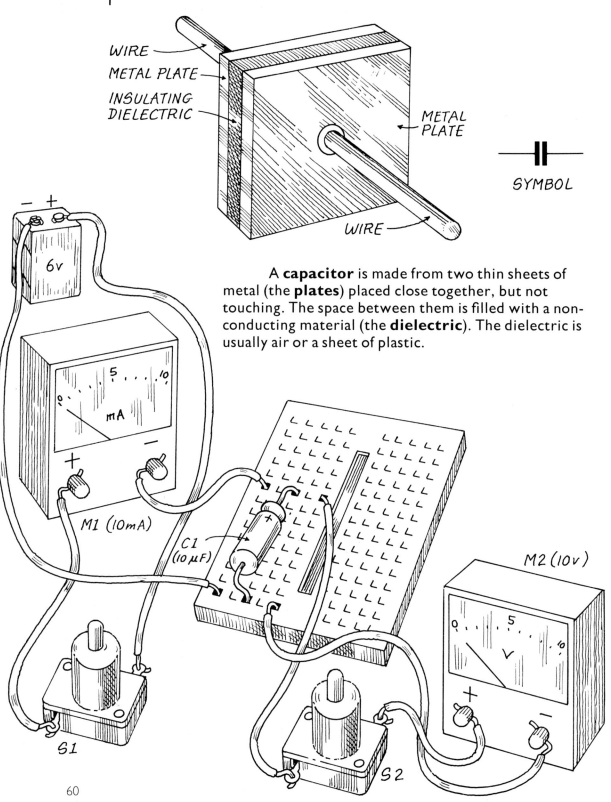

WIRE

METAL PLATE

INSULATING
DIELECTRIC

METAL
PLATE

WIRE

SYMBOL

A **capacitor** is made from two thin sheets of
metal (the **plates**) placed close together, but not
touching. The space between them is filled with a non-
conducting material (the **dielectric**). The dielectric is
usually air or a sheet of plastic.

6v

mA

M1 (10mA)

C1
(10 μF)

M2 (10v)

S1

S2

V

Watch the ammeter M1 while you press and hold S1. The needle kicks, showing that a current flows from the battery to the capacitor for a short time.

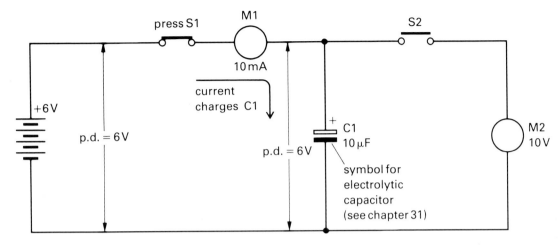

Release S1. The capacitor still has a 6 V p.d. across it. It is **charged** to a p.d. of 6 V.

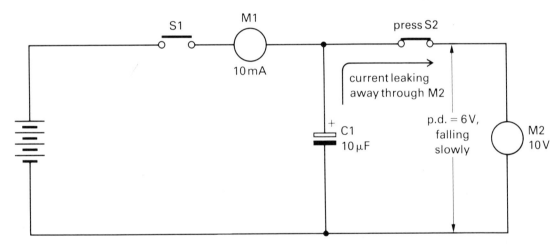

Now press and hold S2. The meter reads 6 V, so the charge is still there, even though the capacitor is not connected to the battery. But, while you hold S2, the p.d. slowly becomes less and less. This is because of the small current which is flowing through the voltmeter. In the end, the p.d. is 0 V, and the capacitor has been **discharged**.

The symbol 'μF' refers to the amount of charge the capacitor can store. The bigger the number, the more it can store. There is more about this in chapter 34.

31 More about capacitors

The test in chapter 30 shows that a capacitor can be used for **storing charge**. It also shows that it takes time for a capacitor to become charged and discharged.

The amount of charge stored depends on:

1   The p.d. between the plates. The bigger the p.d., the more charge is stored.

2   The total area of the plates. The bigger the area, the more charge is stored.

3   The closeness of the plates. The closer the plates, the more charge is stored.

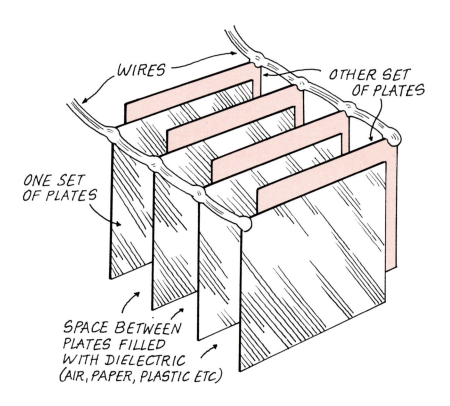

WIRES

OTHER SET OF PLATES

ONE SET OF PLATES

SPACE BETWEEN PLATES FILLED WITH DIELECTRIC (AIR, PAPER, PLASTIC ETC)

Capacitors which store a lot of charge need many large plates that are very close together.

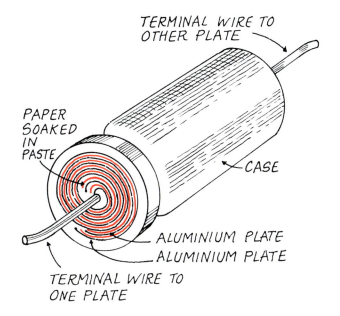

TERMINAL WIRE TO OTHER PLATE

PAPER SOAKED IN PASTE

CASE

ALUMINIUM PLATE
ALUMINIUM PLATE

TERMINAL WIRE TO ONE PLATE

Connect electrolytic capacitors so that the (+) plate is always positive to the (−) plate. If you do not do this, the film of oxide will be destroyed.

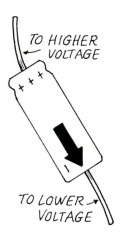

TO HIGHER VOLTAGE

+ + +

TO LOWER VOLTAGE

Electrolytic capacitors are made to hold large amounts of charge. They have aluminium plates of large area, rolled around each other to make the capacitor compact. The plates have a rough surface to give them larger surface area. The space between the plates is filled with paper soaked in a conducting paste. The non-conducting dielectric is a very thin layer of aluminium oxide on the surface of one set of plates. This is made by connecting the capacitor to a power supply. The aluminium oxide is formed by electrolysis.

Other kinds of capacitor

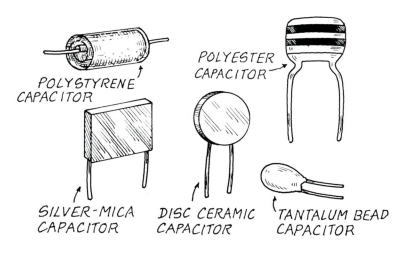

POLYSTYRENE CAPACITOR

POLYESTER CAPACITOR

SILVER-MICA CAPACITOR

DISC CERAMIC CAPACITOR

TANTALUM BEAD CAPACITOR

32 Current and charge

This circuit measures the current flowing out of a capacitor when it is losing its charge. There is a resistor to make the current small, so that the capacitor discharges slowly and you can measure how long it takes.

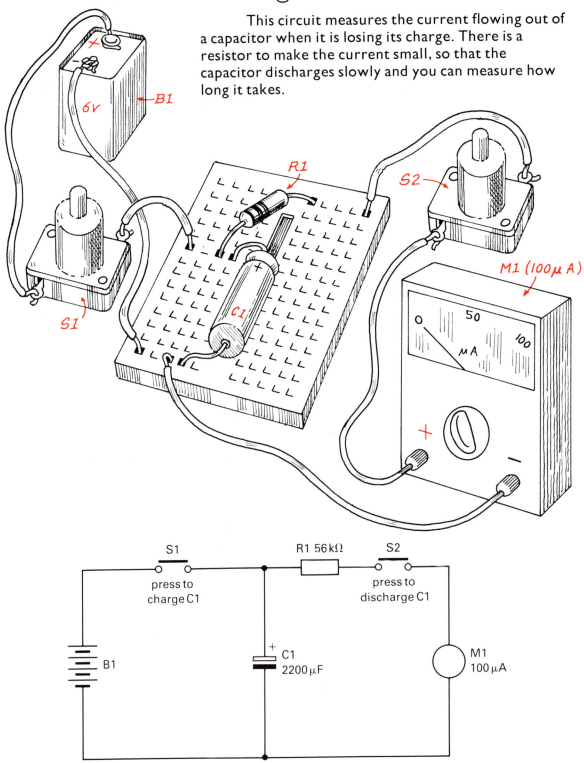

Press S1 and hold it for about a second to charge the capacitor. The capacitor now has a p.d. of 6 V between its plates. If the capacitor loses half of its charge, the p.d. drops to 3 V. The current will then be exactly half of its initial amount (its starting amount). Your partner measures how long it takes for the current to fall from its initial amount (when you first press S2) to half its initial amount, using a watch or clock which shows seconds. Your job is to look at the meter and measure the current.

When your partner says 'Go!', press S2 and hold it. As soon as you press S2 the needle will shoot up to show the initial current. What is **your** initial reading?

Then the reading falls slowly, as the charge flows away through the resistor and meter, and the current becomes less and less. When the current is **half** of its initial amount, say 'Now!'. Your partner will say how many seconds this has taken.

Repeat the measurement a few times to check your result. Then do everything again, using a 220 μF capacitor instead. As you work, fill in a table like this one, **but use your own results**:

VALUE OF CAPACITOR (μF)	INITIAL CURRENT (μA)	FINAL CURRENT (μA)	TIME TAKEN (s)
2200	100	50	90
220	100	50	10

For the figures in the table, the average of the initial and final currents is 75 μA for both capacitors. What was your average current?

33 The coulomb

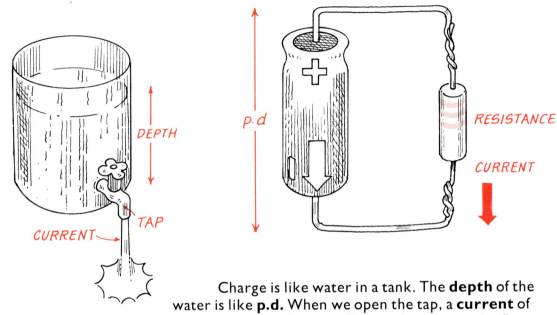

Charge is like water in a tank. The **depth** of the water is like **p.d.** When we open the tap, a **current** of water runs out. The **tap** offers **resistance** to the flow of water.

The deeper the water, the bigger the current.

The more water in the tank, the longer it takes to empty.

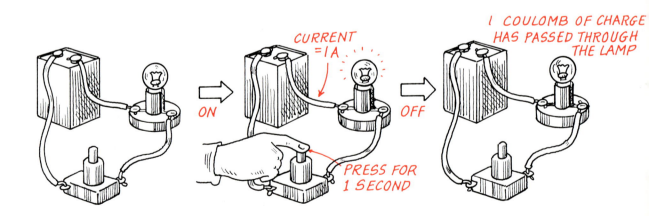

THE CHARGE RULE   If a current of 1 amp flows for 1 second, the amount of charge it carries is 1 coulomb.

CAPACITORS LOSING CHARGE

The charge rule tells us that the bigger the current and the longer the time it flows, the more charge is carried. In the tests on page 65, the initial current and final (half) current were the same for both capacitors. Both capacitors began to discharge at 100 μA. The current fell to 50 μA when the p.d. had fallen to 3 V. At this stage, both capacitors had lost half their charge. But the 2200 μF capacitor took longer to reach this stage. It began with more charge inside it than the 220 μF capacitor, so it took longer to lose half its charge.

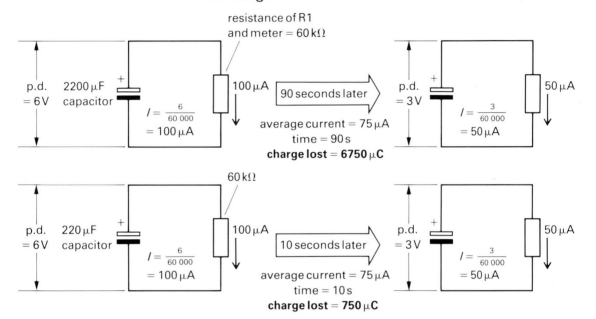

resistance of R1 and meter = 60 kΩ

p.d. = 6 V   2200 μF capacitor   + $I = \dfrac{6}{60\,000}$ = 100 μA   100 μA

90 seconds later

average current = 75 μA
time = 90 s
charge lost = 6750 μC

p.d. = 3 V   + $I = \dfrac{3}{60\,000}$ = 50 μA   50 μA

60 kΩ

p.d. = 6 V   220 μF capacitor   + $I = \dfrac{6}{60\,000}$ = 100 μA   100 μA

10 seconds later

average current = 75 μA
time = 10 s
charge lost = 750 μC

p.d. = 3 V   + $I = \dfrac{3}{60\,000}$ = 50 μA   50 μA

HOW MUCH CHARGE IS LOST?

When a current of 75 μA flows for 90 seconds, the charge it carries away from the capacitor is

current × time = charge
75 μA × 90 s = 6750 μC

The charge is in **microcoulombs** (μC) because we have measured the current in microamps. This is the charge the 2200 μF capacitor loses as its p.d. falls from 6 V to 3 V. The charge lost by the 220 μF capacitor is

current × time = charge
75 μA × 10 s = 750 μC

The 220 μF capacitor loses only 750 μC of charge.

Work out the amounts of charge lost by **your** two capacitors.

34 Capacitance

If we think of a charged capacitor and a tank containing water, these things match:

Tank	matches	**capacitor**
water	matches	charge
current of water	matches	current
depth of water	matches	p.d.
width of tank	matches	?

WHAT MATCHES THE WIDTH OF TANK?

Compare the description on this page with the description on the opposite page.

Here are a wide tank and a narrow one, filled with water to the same depth and with identical taps:

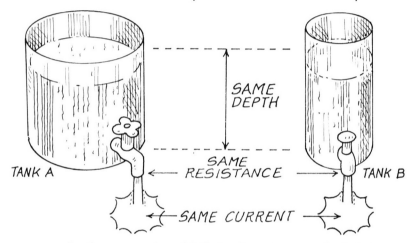

As the water level falls by 1 metre, we find that:

tank A loses 2200 litres of water,
tank B loses 220 litres of water.

This tells us that:

tank A holds 2200 litres of water per metre of depth,
tank B holds 220 litres of water per metre of depth.

The **width** of the tank is matched by the **capacitance** of a capacitor.

If the p.d. across a capacitor rises or falls by 1 volt while it gains or loses 1 coulomb of charge, its capacitance is said to be 1 **farad** (F).

In short:

THE CAPACITANCE RULE

capacitance = charge gained or lost ÷ p.d.
    (F)   (C)   (V)

WORKING OUT CAPACITANCE

We can work out the actual capacitances of the two capacitors tested on page 65. These values will not be exactly the same as the values printed on each capacitor, but should be fairly close.

We use the capacitance rule and the values of 'charge lost' worked out on page 67:

For the '2200 μF' capacitor:

capacitance = 6750 μC ÷ 3 V = 2250 μF

For the '220 μF' capacitor:

capacitance = 750 μC ÷ 3 V = 250 μF

These results are very close to the capacitances marked on each capacitor. Now work out the capacitances of the capacitors you used.

Here are a 2200 μF capacitor and a 220 μF capacitor, charged to the same p.d., and connected to identical resistors.

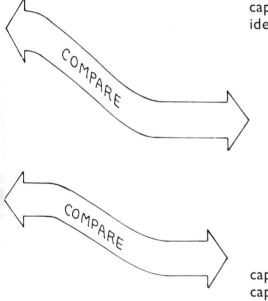
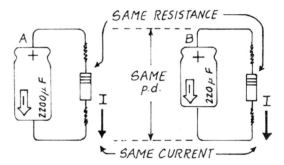

As the p.d. falls by 1 volt, we find that:

capacitor A loses 2200 microcoulombs,
capacitor B loses 220 microcoulombs.

This tells us that:

capacitor A holds 2200 microcoulombs per volt,
capacitor B holds 220 microcoulombs per volt.

PRACTICAL UNITS

The farad is much too large a unit to be used in practical electronics. We use these units instead:

microfarad (μF) = millionth of a farad

nanofarad (nF) = thousand-millionth of a farad

picofarad (pF) = million-millionth of a farad

35 Summing up 2

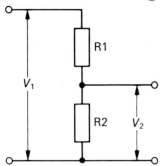

In the **potential divider** on the left, the p.d. across R2 is calculated from:

$$V_2 = V_1 \times \frac{R_2}{R_1 + R_2}$$

SEMICONDUCTOR DEVICES

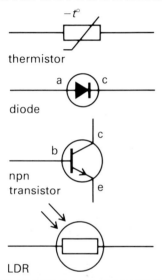

thermistor

diode

npn transistor

LDR

A **thermistor's** resistance becomes less as it gets warmer.

A **diode** conducts in only one direction, from p-type to n-type (or from anode wire to cathode wire).

In an **npn transistor** (p-type base layer between collector and emitter layers of n-type), a small current flowing into the base causes a much larger current to flow into the collector and out of the emitter.

A **light-dependent resistor's** resistance becomes less when more light shines on it.

INPUT AND OUTPUT OF A TRANSISTOR

When the input is 0 V, the output is 6 V. When the input is 6 V, the output is 0 V.

BISTABLE CIRCUITS

A circuit with two stable states (e.g. a flip-flop) is called a bistable circuit.

CAPACITORS

A simple capacitor consists of two metal plates separated by an insulating dielectric. It can be used to store charge. The amount of charge stored is bigger if (a) the p.d. is large, (b) the plates are large, and (c) the plates are close together.

THE CHARGE RULE

Charge = time × current
(coulombs)   (seconds)   (amps)

THE CAPACITANCE RULE

Capacitance = change in charge ÷ change in p.d.
(farads)   (coulombs)   (volts)

1   What do we mean by the 'rule of opposites'?

2   Why is a flip-flop called a bistable circuit?

3   Suppose you are shown a flip-flop like the one on page 56, which someone else has been working with. They have left the power on but the wire D is left disconnected. D1 is on and D2 is off. What was the last thing the person did to the circuit before you saw it? What must you do now to reverse the state of the LEDs?

4   In what part of a computer are you likely to find a lot of flip-flops?

5   What are the main parts of a capacitor?

6   How can a capacitor be made to store a lot of charge?

7   Name a type of capacitor which is often used for storing large amounts of charge.

8   A capacitor is charged with a steady current of 200 μA for 30 seconds. How much charge does it receive?

9   A capacitor holds a charge of 125 mC. A steady current of 2 mA is flowing from it. How long does it take to reduce the charge to 100 mC?

10   A 100 μF capacitor is charged to a p.d. of 6 V. How much charge does it hold? It is then charged to 9 V. How much extra charge does it now hold? The capacitor is then discharged through a 270 Ω resistor. What is the current as it begins to discharge? What is the current as it finishes discharging?

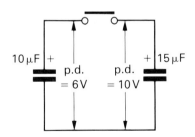

11   The two capacitors in the drawing are each charged as shown. How much charge does each hold? In which direction does the current flow when S1 is pressed? If S1 is pressed and held until no more current flows, what will be the final p.d. across the capacitors?

36 A delay circuit

Compare the circuit below with the flip-flop shown on page 59. What differences do you notice?

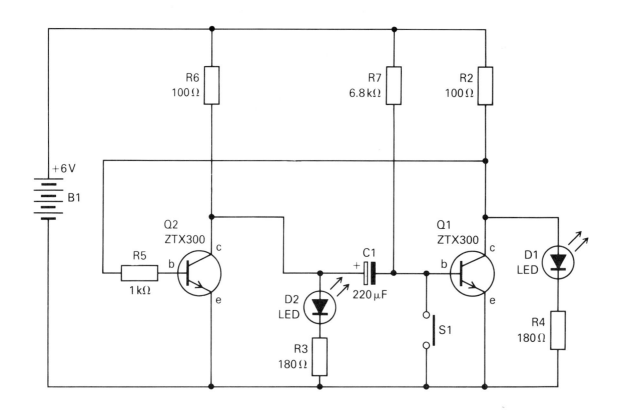

The main difference is that, instead of the resistor R1, we now have a capacitor C1. There is an extra resistor R7 and there is only one button-switch (S1).

Connections in a flip-flop
In a flip-flop, the output from each transistor is connected to the input of the other transistor through a **resistor** (R1 and R5 on page 59).

Connections in this circuit
In this circuit, the output of Q1 is connected to the input of Q2 by a **resistor** (R5), **but** the output of Q2 is connected to the input of Q1 by a **capacitor** (C1).

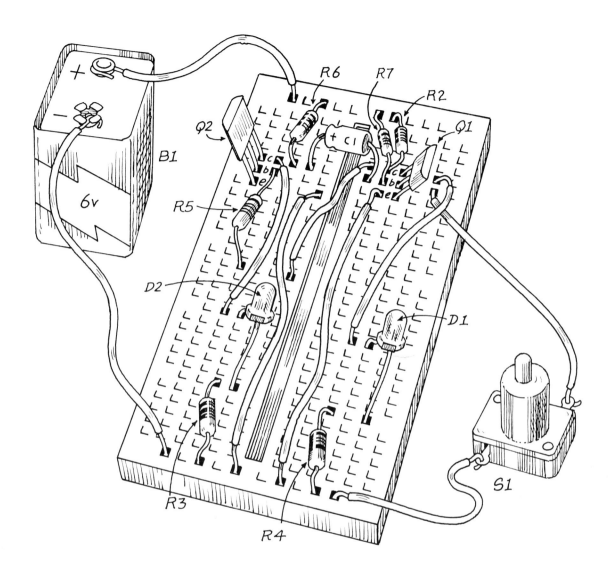

When you connect the battery, D1 is off (showing that Q1 is on) and D2 is on (showing that Q2 is off). It is like the top drawing on page 59.

Press S1, but only for an instant. D1 comes on and D2 goes off. The circuit has 'flipped'. Now wait – what happens next?

After a second or two, the circuit 'flops' back to its first state again. It is stable in its first state, but it is unstable in its second state. After a short time in its unstable state, it goes back to the stable state and stays there.

37 One stable state

In the delay circuit of chapter 36, the reason for the delay is the capacitor C1. In the stable state (below), a small current flows through the high resistance R7. The base of Q1 is at about 0.7 V. The current is enough to turn Q1 on. There is a p.d. of about 5.3 V across C1.

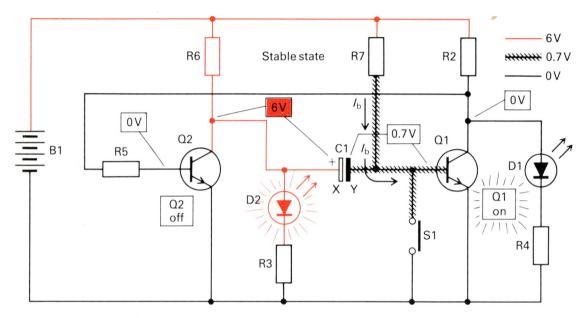

When you press S1, Q1 turns off, Q2 turns on and the voltage at plate X of C1 very suddenly falls to 0 V. But C1 cannot lose its charge quickly. The p.d. across C1 does not change quickly.

When the voltage at plate X drops from 6 V to 0 V, the voltage at plate Y drops by the same amount, from 0.7 V to −5.3 V.

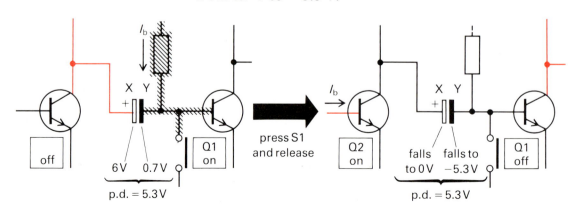

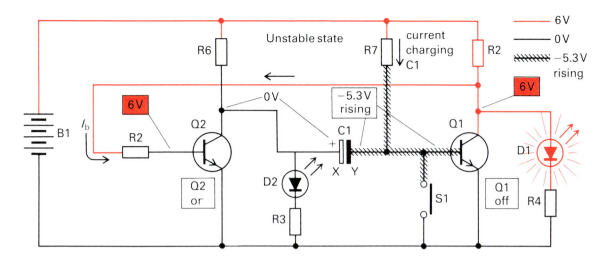

In the unstable state (above), plate Y is at -5.3 V to begin with. Current flows through R7, gradually charging plate Y. The voltage at plate Y gradually rises, until it comes to about 0.7 V. This is just enough to turn Q1 on. When Q1 turns on, Q2 turns off, sending the circuit back to its stable state.

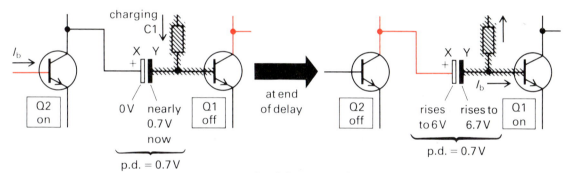

As Q2 turns off, the voltage at plate X rises quickly from 0 V to $+6$ V. The p.d. between X and Y cannot change quickly. Plate Y rises from 0.7 V to $+6.7$ V. Then the charge on Y leaks away through Q1 until the voltage at Y has fallen to 0.7 V again. The small current flowing through R7 is enough to keep Q1 turned on. The circuit is stable.

The circuit is stable in only **one** state, so we call it a **mono**stable circuit. The length of time it remains in its unstable state depends on the capacitance of C1 and the resistance of R7.

Measure the delay of your circuit, in seconds. Now change C1 for a 100 μF capacitor, and then for a capacitor of larger value, say 2200 μF. Explain why the length of delay is altered.

38 An oscillating circuit

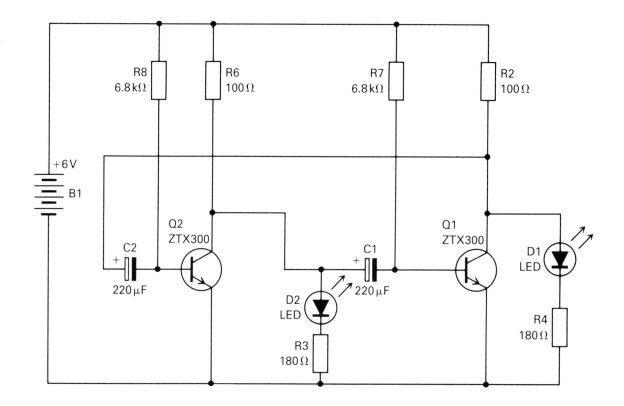

Here is another circuit which looks like the flip-flop of chapters 28 and 29. It also looks like the monostable circuit of chapters 36 and 37. This circuit has two capacitors. One capacitor (C2) joins the output of Q1 to the input of Q2. The other capacitor (C1) joins the output of Q2 to the input of Q1. As in the monostable circuit, each capacitor has a resistor joined to it, through which current can flow to charge it or discharge it.

The capacitor in a monostable circuit makes the circuit unstable in **one** of its states. After a short time in the unstable state, it returns to the stable state. What do you think happens in this circuit, which has **two** capacitors? Build the circuit and find out.

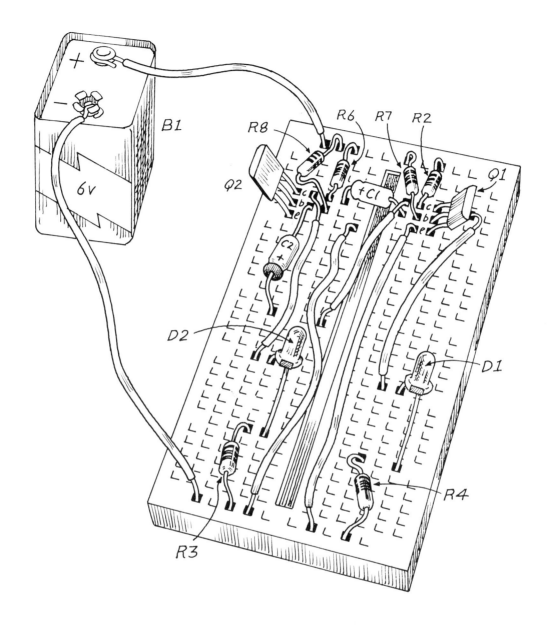

When you connect the power supply, the LEDs immediately begin to flash on and off regularly. This circuit is unstable in **both** states. After a short time in one unstable state, it changes to the other unstable state. It keeps on changing state – it **oscillates**.

Keep this circuit made up for chapter 39.

39 Frequency

Count the number of flashes made by one of the LEDs during a fixed time, such as 30 seconds. Divide the number by 30. This tells you how many times it flashes per second.

The number of flashes each second (from one of the LEDs) is called the **frequency** of the oscillator. If it flashes once every second, the frequency is 1 per second, or 1 **hertz** (1 Hz). If it flashes 4 times in a second, the frequency is 4 Hz. What is the frequency of your oscillator?

HOW THE OSCILLATOR WORKS

It changes backward and forward from one unstable state to the other.

A circuit which has no stable state is called an astable circuit.

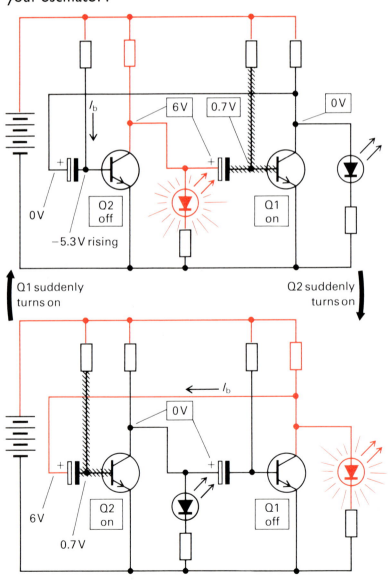

Charge is like water in a tank (see page 66).

The level of water in a tank falls quickly if:

**(a) the tank is narrow, or
(b) the tap is wide, or
(c) the tank is narrow and the tap is wide.**

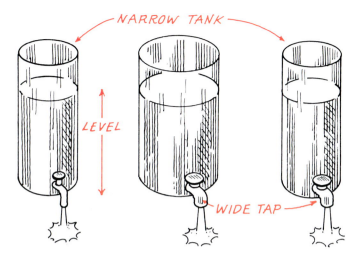

The level of water in a tank falls slowly if:

**(a) the tank is wide, or
(b) the tap is narrow, or
(c) the tank is wide and the tap is narrow.**

Capacitors behave in a similar way.

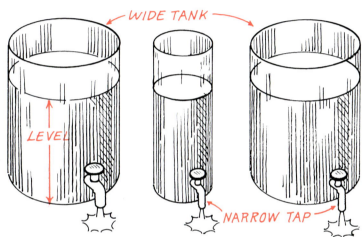

ALTERING FREQUENCY

The level of charge on a capacitor changes more quickly if it has low capacitance **or** if current can leave it easily (charging or discharging through a low-value resistor). It changes more slowly if the capacitance and resistance are high. The frequency of the oscillator depends on how fast the capacitors charge and discharge. So if you alter the values of C1, C2, R7 and R8, this should alter the frequency.

Increase C1 and C2 to 470 μF or 1000 μF each. Find the new frequency. Has the frequency increased or decreased?

Put back the 220 μF capacitors. Increase R7 and R8 to 18 kΩ each. Has the frequency increased or decreased?

Decrease C1 and C2 to 47 μF. What is the frequency now?

Keep this circuit made up for chapter 40.

An oscillator which produces frequencies we can hear is called an audio-frequency oscillator.

40 High-frequency oscillator

Make up the oscillator circuit from chapters **38** and **39** (see page 76). Use 100 nF (= 0.1 μF) capacitors for C1 and C2. Connect the battery and watch the LEDs.

The capacitance of C1 and C2 is low, so they are charged and discharged quickly. This means that the oscillator has a higher frequency than it had before. The LEDs flash on and off so quickly that you cannot see the flashes. They seem to be on all the time, but shining about half as brightly.

The LEDs are no use for showing the outputs of the oscillator now. So unplug them and their resistors (R3, R4).

Connect a loudspeaker to the output of Q1. The reason for having the third capacitor (C3) is explained later (chapter **41**), but do not bother about this at present. When the circuit is ready, connect it to the battery.

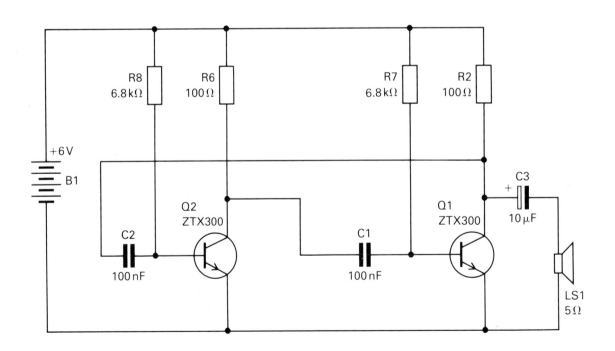

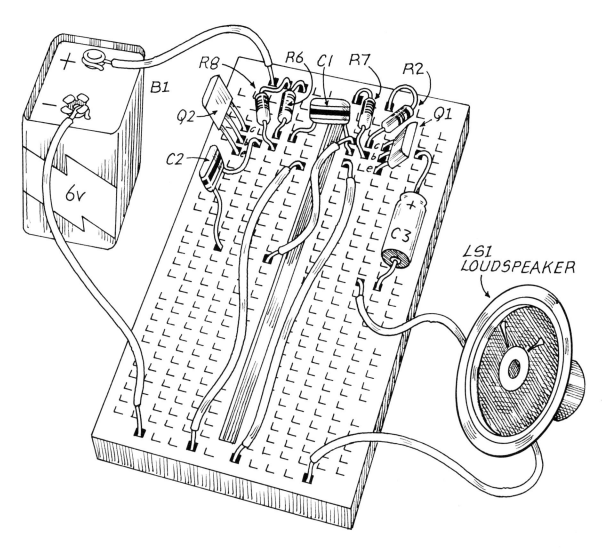

The circuit oscillates as explained on page 78. As Q1 switches off, a current flows through the coil of the loudspeaker (see the next chapter), generating a magnetic field around it. The coil is attracted toward the permanent magnet of the loudspeaker. The coil is attached to the cone of the loudspeaker, so the cone moves too. Then, when Q1 switches on, the magnetic field disappears and the cone moves back again. If Q1 switches on and off several hundred times a second (a frequency of several hundred hertz), the cone moves backward and forward at the same frequency. This motion causes the air around the cone to vibrate. If the frequency is between about 30 Hz and 15 kHz, we sense these vibrations as sound.

Keep the circuit made up for chapter 41.

41 Oscillators and sound

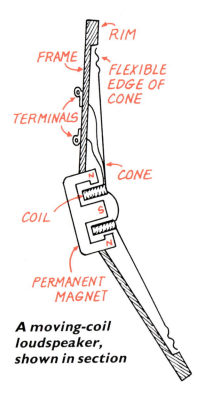

A loudspeaker has a coil, which is held in place between the poles of a permanent magnet. The coil is attached to the cone of the loudspeaker. The cone is attached to the frame of the loudspeaker by a flexible region which allows the cone to move slightly.

When a current from the oscillator passes through the coil, a magnetic field appears around the coil. This reacts with the magnetic field of the permanent magnet. The result is a force which makes the coil move. Since the coil is attached to the cone, the cone moves too. The picture on the opposite page shows that movement of the cone causes movement of the air. The moving cone vibrates, causing vibrations (or sound waves) to spread through the air. We hear the sound when these waves reach our ears.

A moving-coil loudspeaker, shown in section

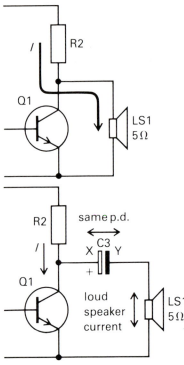

The coil of the loudspeaker has very low resistance (only about 5 Ω). If it was connected directly to the collector of Q1 (see left), nearly all the current coming through R2 would go through the loudspeaker and very little would go through Q1. The oscillator would not work.

When C3 is wired between Q1 and LS1, the current cannot by-pass Q1. The oscillator works properly, and the voltage at the collector rises and falls at audio frequency. As the voltage at plate X rises, the p.d. between X and Y does not change (see page 74). So the voltage at Y rises. An instant later the voltage at X falls, making the voltage at Y fall too. The voltage at Y rises and falls at the frequency of the oscillator. The rising and falling voltage at Y causes a current to flow through the coil of the loudspeaker, first in one direction and then in the other. Sound is generated, as explained above.

In this way C3 is used to **couple** the loudspeaker to the oscillator circuits.

How we hear the sounds

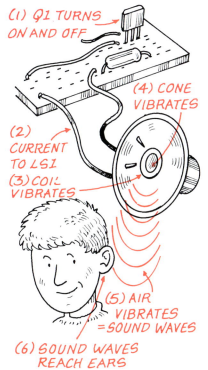

(1) Q1 TURNS ON AND OFF

(2) CURRENT TO LS1

(3) COIL VIBRATES

(4) CONE VIBRATES

(5) AIR VIBRATES = SOUND WAVES

(6) SOUND WAVES REACH EARS

Do not hold the plates, or your body will affect the capacitance.

Change C1 and C2 to alter the frequency of the sound. If C1 and C2 have greater capacitance (470 nF), the frequency is reduced. The pitch of the sound is lowered. If C1 and C2 have lesser capacitance (47nF), the frequency is increased. The pitch is raised.

Put back the 100 nF capacitors, and change R7 and R8 for 100 kΩ resistors. With greater resistance, the capacitors take longer to charge and discharge. The frequency (and therefore the pitch) is lowered.

MAKE YOUR OWN CAPACITOR

Replace C1 with a home-made capacitor. You can alter its capacitance in three ways.

1 First make the capacitor with large plates (30 cm $\times$ 30 cm) and listen to the note it produces. Then cut the plates down to 5 cm $\times$ 5 cm and listen again.

2 Lift up the paper, so that the distance between the plates is increased. Listen to the effect of reducing capacitance in this way.

3 Slide the top plate to one side, so the area of overlap is reduced.

Also try the effect of using different dielectrics, such as a sheet of glass, or a sheet of plastic.

What happens when there is no dielectric and the sheets come into contact? If the plates are in contact, capacitance is zero.

Try playing a tune with your variable capacitor!

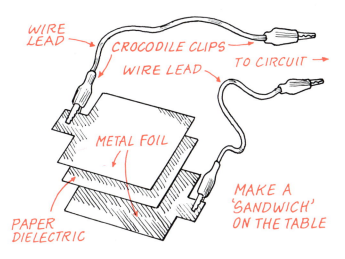

WIRE LEAD

CROCODILE CLIPS

WIRE LEAD

TO CIRCUIT

METAL FOIL

MAKE A 'SANDWICH' ON THE TABLE

PAPER DIELECTRIC

42 Pulses and waves

If a monostable circuit is working from a 6 V battery, the output from one of the transistors changes from 0 V to 6 V when the circuit is triggered. After a fixed time, the circuit returns to its stable state and the output falls from 6 V to 0 V again. We could illustrate these changes by drawing a graph, like this:

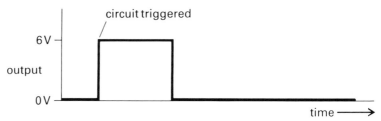

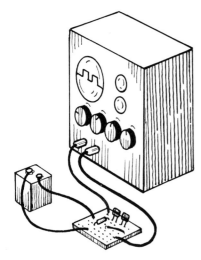

The best way to see what the waveforms look like is to connect the output of the circuit to an oscilloscope.

The graph shows one **pulse** from the monostable circuit. The output from the other transistor of the monostable circuit looks like this.

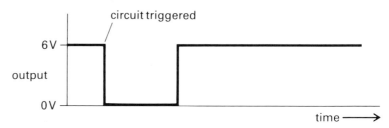

This pulse starts with a fall in voltage. We call it a **negative-going pulse**, because the voltage is falling in the negative direction.

The output of an astable circuit changes from 0 V to 6 V and back to 0 V over and over again, so the graph looks like this.

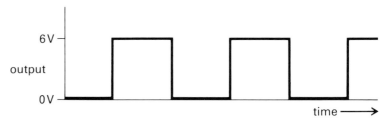

It takes time for the output to rise or fall, so the waves are not perfectly square, but slightly rounded.

The graph shows a **series of pulses**. It looks like a series of waves. We call the graph the **waveform** of the output. The voltage changes very quickly so there is a sharp rise and fall. These are called **square waves**. The number of pulses per second is the frequency of the wave. It is given in hertz.

A monostable circuit may be triggered by briefly connecting the base of one of the transistors to 0 V. The negative-going input pulse and the output pulse which it causes can both be drawn on the same graph.

This is the waveform of the triggering pulse.

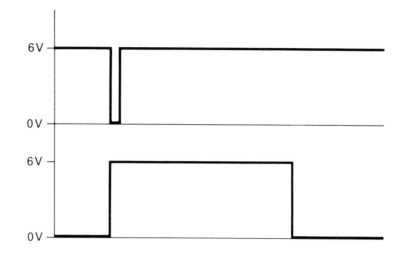

This is the waveform of the output.

The graph shows that the output pulse is much longer than the input pulse.

Some kinds of oscillating (or astable) circuits give different waveforms.

Sawtooth waveform

Triangular waveform

Sine waveform

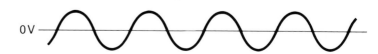

43 D.C. and A.C.

A steady (or fairly steady) current, which always flows in the same direction, is called **direct current** (or d.c. for short). This is the kind of current we get from a battery.

If a current reverses in direction regularly, it is called **alternating current** (or a.c. for short). This is the kind of current we get from a mains socket.

This graph shows the alternating current flowing through a lamp connected to the mains.

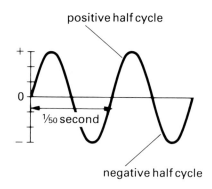

positive half cycle

negative half cycle

¹⁄₅₀ second

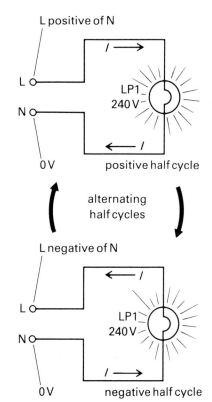

A mains socket is connected to three wires or lines, called earth, live and neutral.

The earth (E) line is connected to the earth somewhere in your home. It is often connected to a metal pipe of the cold-water system. It is at 0 V (earth potential).

The neutral (N) line is connected to the earth at the power generating station, so it is at 0 V too. One terminal of the generator is connected to this line.

The live line (L) is connected to the other terminal of the generator. The output from this terminal is a sine wave, with a frequency of 50 Hz. In one-fiftieth of a second, the output swings from zero to positive, then back to zero, then swings to negative and back to zero again.

When a lamp is connected to the mains (see the diagrams on the left), current flows from the live line through the lamp to the neutral line when L is positive to N. In the other half of the cycle, the current flows from the N line into the L line.

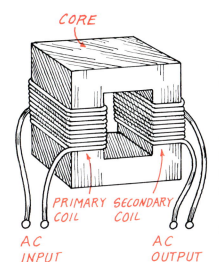

CORE

PRIMARY COIL   SECONDARY COIL

AC INPUT   AC OUTPUT

Mains current is not suitable for transistor circuits. The voltage (240 V) is too high for most circuits. Transistors need d.c.

We can reduce the voltage by using a **transformer**. This consists of two coils of wire wound on the same soft iron core. If the secondary coil has fewer turns than the primary coil, the voltage at the secondary coil is lower than 240 V. But the current is still a.c.

We use diodes to turn a.c. into d.c. This is called **rectification**. Here are two ways of doing it.

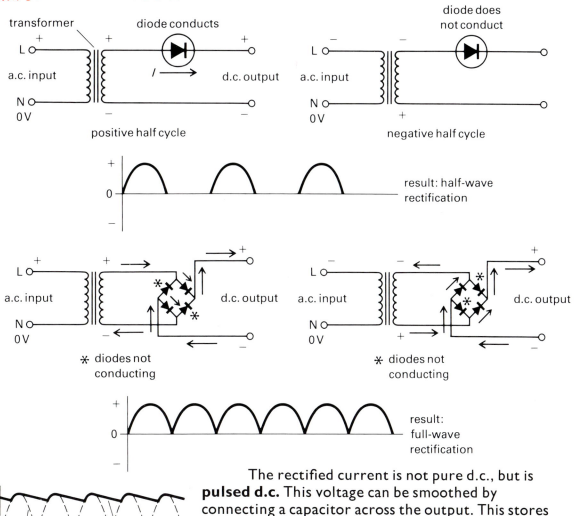

transformer   diode conducts
+   +
L   +
a.c. input   I →   d.c. output
N
0V   −   −
positive half cycle

diode does not conduct
−   −
L
a.c. input   d.c. output
N
0V   +
negative half cycle

result: half-wave rectification

+
L
a.c. input   →   +
N   →   d.c. output
0V
− ←   −
* diodes not conducting

−
L
a.c. input   ←   +
N   →   d.c. output
0V   +   −
* diodes not conducting

result: full-wave rectification

pulsed d.c.   d.c. smoothed by capacitor

The rectified current is not pure d.c., but is **pulsed d.c.** This voltage can be smoothed by connecting a capacitor across the output. This stores charge when the voltage is at its peak, and releases it when the voltage falls, between pulses. In this way the circuit gets a fairly steady supply of current.

44 Relays

A relay is an electrically operated switch. Usually a relay consists of a **coil** wound round a soft iron **core**, a moving **armature**, and a pair of **switch contacts**. The armature is held in position by the springiness of the switch contacts.

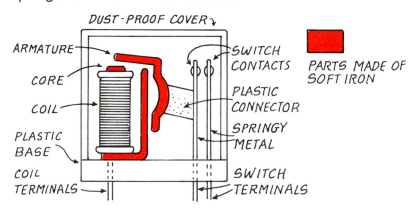

Some relays have more than one set of contacts. They can switch lots of circuits at once. When the armature moves, some pairs of contacts close and others open. Some circuits are turned on and others are turned off.

The coil is wired into a circuit which we will call the first circuit. The coil and its core make up an electromagnet. When S1 is closed, a current passes through the coil, and the core becomes magnetised. The magnetised core attracts the armature, which is usually made of soft iron. The armature is able to turn so that one end of it moves toward the magnetised core. The other end of the armature moves too, pressing the two contacts together. When the contacts touch, current flows in the second circuit.

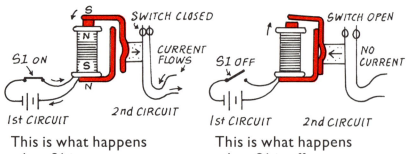

This is what happens when S1 is on.

This is what happens when S1 is off.

Soft iron loses its magnetism easily. When the current in the first circuit is switched off, the core is no longer magnetised. The armature is released and the springy contact pushes it back to its original position. The contacts are no longer held together and the second circuit is broken.

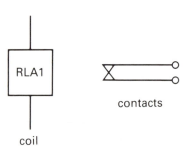

The symbols of a relay coil and the contacts.

WHY USE A RELAY?

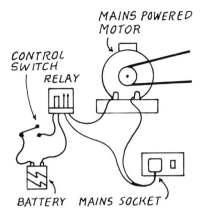

Switching a motor with a relay

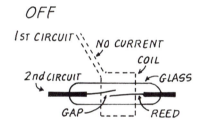

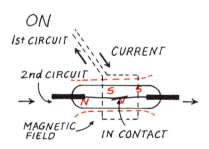

The parts of a reed relay

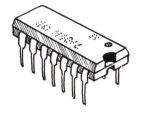

Most electromagnetic relays need only a low voltage (6 V, 12 V or 24 V) for the coil circuit and use a small current. But the switch contacts can carry large currents (many amps) and work at high voltages (up to, say, 240 V). Such a relay is a device for switching a large current at high voltage by using a small current at low voltage.

Relays can switch alternating currents such as the a.c. mains just as easily as direct currents. Relays are used to control lights in theatres, the motors in lifts, and in many other ways.

REED RELAYS

This is another type of relay, which uses a smaller current and switches smaller currents than the previous type. It takes up much less room and works at much higher speed than the other type. As before, we use it when we do not want the first circuit to be electrically connected to the second circuit. For example, we might use it to make a remote-control circuit switch the circuits in a TV set. We could not want the remote-control circuit (the first circuit) to be actually joined to the TV circuit (the second circuit) for this might upset the working of the TV set. For example, the second circuit may be carrying alternating current, such as sound or video signals.

A reed relay has two springy contacts (reeds), sealed in a glass bulb to keep out dust. The bulb is surrounded by a coil. When a current flows in the first circuit a magnetic field is formed by the coil. The reeds become magnetised, as shown on the left.

They are attracted to each other, come together and make contact. Now current can flow in the second circuit.

When the current in the first circuit is switched off, the reeds lose their magnetism and are not attracted together. They spring apart and the second circuit is broken.

A reed relay without a coil can also be switched on by placing a permanent magnet near it.

Reed relays are usually small. The integrated circuit package on the left contains a reed relay. It can be mounted on a circuit board along with other components.

45 Summing up 3

USING TRANSFORMERS   A transformer converts a.c. at one voltage to a.c. at another voltage (which may be higher, lower or the same).

USING DIODES   We use diodes to rectify a.c. to make it into a kind of d.c. suitable for use in electronic circuits.

USING CAPACITORS   We use capacitors to transfer sudden changes of voltage from one part of a circuit to another. If the voltage on one plate is made to change quickly, the p.d. between plates does not change quickly, and so the voltage of the other plate is forced to change too.

USING TRANSISTORS   We can use transistors in switching circuits. If two such circuits are linked together, so that the output of each one is the output of the other, we can build three different kinds of circuit:

1   Resistor-resistor links: gives a bistable circuit (two stable states), often called a flip-flop, and used in memory circuits of computers.

2   Resistor-capacitor links: gives a monostable circuit (one stable state), used to produce a delay.

3   Capacitor-capacitor links: gives an astable circuit (no stable state) used as an oscillator.

USING RELAYS   We can use a transistor for switching a large current by means of a small one, but we cannot switch an alternating current in this way. For switching a.c. we use electromagnetic relays. These are used for switching large or small a.c., using small d.c. For switching small currents we can use a reed relay, a simpler type of electromagnetic relay.

FREQUENCY   The unit of frequency is the hertz. 1 Hz is a frequency of 1 oscillation per second. The range 30 Hz to about 15 kHz is the frequency range of sound (audio frequencies).

REVISION QUESTIONS

1 How many stable states has (a) a bistable circuit, (b) an astable circuit, and (c) a monostable circuit?

2 In a flip-flop like that on page 56, each transistor is joined to the other one by a resistor. How can you convert this circuit into (a) a monostable circuit, and (b) an astable circuit?

3 What kind of circuit could you use to produce a time delay? If you wanted to increase the length of the delay, what change could you make to the circuit?

4 If a lamp flashes with a frequency of 45 Hz, how often does it flash in 10 seconds?

5 What do we call an oscillator which works with a frequency in the range 30 Hz to about 15 kHz?

6 What is the frequency of the mains supply?

7 Which two lines of the mains supply are connected to the generator? Which two are connected to earth?

8 What names are given to the waves shown in the diagrams below?

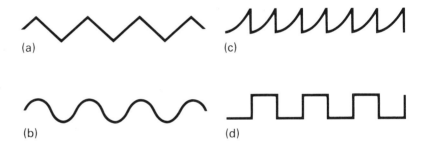

(a)   (c)

(b)   (d)

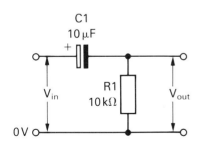

9 In the part-circuit on the left, $V_{in} = +2\ V$, and $V_{out} = 0\ V$. If V_{in} is suddenly increased to $+5\ V$, what voltage does V_{out} have at first? What current will begin to flow through R1? If V_{in} remains steady at $+5\ V$, how much additional charge will have been stored in the capacitor?

10 Design an audio-frequency oscillator circuit, the frequency of which depends on the brightness of the light in the room the oscillator is in.

11 Design a circuit for use in a greenhouse, to switch on a mains-powered fan when the temperature in the greenhouse becomes too hot.

12 Design a circuit to switch on a porch-light for a few seconds whenever a button beside the door is pushed.

List of components required

The following items are needed by each individual or group building the circuits in this book:

Quantity Description

1
Breadboard, 2.5mm pitch. This can be of any size. The Verobloc is suitable for all circuits described in this book.

1
Power supply, giving 6 V d.c. (unregulated, and up to about 150 mA), and tapped at 3 V. This can be a mains-powered pack, a battery holder with four dry cells ('D' type preferred) or an array of Nicad or NiFe cells in series.

2
Button-switches (the push-to-**make** contact kind) with bare-ended leads attached.

8
8 cm lengths of 1/0.6 mm tinned copper wire, with p.v.c. insulation, stripped for about 10 mm at both ends.

2
15 cm lengths of wire, as above, stripped at one end, and with an insulated crocodile clip at the other.

1
Multimeter, with the following f.s.d. direct-current ranges: 100 μA, 10 mA, 100 mA, 10 V, and the usual resistance ranges. Separate ammeters and voltmeters may be used instead, and for two tests (chapters 12 and 30) an ammeter and a 10 V meter are needed at the same time.

2
Filament lamps, 6 V, 60 mA, in sockets; the sockets to have bare-ended leads attached. Alternative: wire-ended lamps of the same specification.

2
ZTX300, npn transistors, or similar general-purpose transistors in E-line case.

1
1N4001, 1N4148 or other silicon diode.

2
TIL209 or similar light-emitting diode, preferably red.

1	VA1026 thermistor
1	ORP12 or similar light-dependent resistor, with 8 cm wire leads soldered to each terminal, insulation stripped at their other ends.
1 each	of the following values of carbon resistor, 0.125 W, 0.33 W or 0.25 W, tolerance 5% or 10%: 56 Ω, 82 Ω, 220 Ω, 270 Ω, 330 Ω, 390 Ω, 470Ω, 560Ω, 2.7kΩ, 3.3Ω and 56 kΩ.
2 each	of the following values of resistor as above: 100 Ω, 180 Ω, 1 kΩ, 6.8 kΩ and 18 kΩ.
1 each	of the following electrolytic capacitors (10 V): 10 μF and 100 μF.
2 each	of the following capacitors: 100 nF (polyester), 220 μF (electrolytic, 10 V), 2200 μF (electrolytic, 10 V). In addition, 2 of each of the following capacitors are to be shared between two or more groups: 47 nF (polyester), 470 nF (polyester), 47 μF (electrolytic, 10 V), and 470 μF or 1000 μF (electrolytic, 10 V).
1	miniature loudspeaker, 5 ohm (3 ohm or 8 ohm can also be used), with 8 cm leads soldered to the terminals, their other ends being stripped.

An oscilloscope is useful for demonstrating the waveforms in chapter 42. It is helpful if a simple electromagnetic relay and a reed relay can be available for examination in chapter 44.

Addresses of suppliers
(all will supply by mail order)

Breadboard:
BICC-Verospeed, Stansted Road, Boyatt Wood, Eastleigh, Hants, SO5 4ZY. (0703–618525) (for the Verobloc).

Other components (including Verobloc and other breadboards):
Maplin Electronic Supplies Ltd, PO Box 3, Rayleigh, Essex, SS6 8LR. (0702–554155) (Their mail order catalogue is on sale at most branches of W.H. Smith.).

Electrovalue Ltd, 28 St. Jude's Road, Englefield Green, Egham, Surrey, TW20 0HB. (0784—33603).

Watford Electronics, 35 Cardiff Road, Watford, Herts., WD1 8ED. (0923–40588).

Magenta Electronics Ltd, 135 Hunter Street, Burton-on-Trent, Staffs., DE14 2ST. (0283–65435).

Schools and other educational establishments may obtain components from: RS Components Ltd, PO Box 99, Corby, Northants NN17 9RS, U.K.